Pests Control and Acarology

Edited by Dalila Haouas and Levente Hufnagel

Published in London, United Kingdom

IntechOpen

Supporting open minds since 2005

Pests Control and Acarology
http://dx.doi.org/10.5772/intechopen.76737
Edited by Dalila Haouas and Levente Hufnagel

Contributors
Hamadttu El-Shafie, Loth S. Mulungu, Bernard M. Mchukya, Laurent L. Mnyone, Talha Nazir, Sehroon Khan, Dewen Qiu, Abou Togola, Ousmane Boukar, Manuele Tamo, Siva Chamarthi, Carlos Vásquez, Yelitza Colmenarez, Muhammad Sarwar, Sigmund Hågvar, Levente Hufnagel, Ferenc Mics, Réka Homoródi, Merlin Isaac Kamala

First published in London, United Kingdom, 2020 by IntechOpen
IntechOpen is the global imprint of INTECHOPEN LIMITED, registered in England and Wales, registration number: 11086078, 7th floor, 10 Lower Thames Street, London, EC3R 6AF, United Kingdom
Printed in Croatia

British Library Cataloguing-in-Publication Data
A catalogue record for this book is available from the British Library

Additional hard and PDF copies can be obtained from orders@intechopen.com

Pests Control and Acarology
Edited by Dalila Haouas and Levente Hufnagel
p. cm.
Print ISBN 978-1-83880-602-6
Online ISBN 978-1-83880-603-3
eBook (PDF) ISBN 978-1-83880-604-0

For EU product safety concerns:
IN TECH d.o.o., Prolaz Marije Krucifikse Kozulić 3, 51000 Rijeka, Croatia, info@intechopen.com or visit our website at intechopen.com.

Meet the editors

Dr. Dalila Haouas is an assistant professor at the Graduate Agriculture School of Kef, University of Jendouba, Tunisia. She teaches Agricultural Entomology and Zoology. Her research interests include invasive and emergent insects on crops, botanical extracts and products for insect pest management. She has published articles in many international journals such as *Industrial Crops and Products*, *Journal of Pest Science*, *Chemistry and Biodiversity*, *Bulletin OEPP/EPPO Bulletin*, and *Shilap-Revista De Lepidopterologia*. She has also presented many papers at conferences, both locally and abroad. She is a reviewer of articles and chapters in impacted and indexed international journals.

Dr. Levente Hufnagel is an associate professor at Szent István University, Gödöllő, Hungary, in climate change and ecosystem ecology, biogeography, biometrics, and ecological research methodology. He has over 20 years of experience leading Hungarian academic institutions teaching PhD, MSc, and BSc students from various social and cultural backgrounds. He has more than 160 scientific publications (in both aquatic and terrestrial ecological aspects of plants, animals, and microbes, at community as well as population level) and more than 700 independent citations. As a participant of several big ecological research and development projects, he has significant experience in multidisciplinary cooperations (with more than 180 coauthors in different publications). As the supervisor of several PhD, BSc, and MSc theses, and as editor-in-chief of an international scientific journal indexed by Web of Science and Scopus, he has broad supervising and editing experience. Dr Hufnagel graduated from Eötvös Lorand University with a Master degree in Ecology and Evolutionary Biology and a PhD degree in Hydrobiology. He also graduated from Szent István University with a PhD degree in Agricultural Science.

Contents

Preface

The imperatives of productivity, which impose high yields of quality plant products combined with the specialization of crops by region, make plant protection a vital activity in agriculture. Pest control has made huge progress during the 20th century. This progress has been made possible by scientific and technical breakthroughs, particularly in chemistry (analytical and synthetic) and in biology (population dynamics, ecosystem analysis, biological control theory and practice, biotechnology). For these reasons, and to ensure food safety, several methods of protection are used. Indeed, after the Second World War, the appearance of inexpensive synthetic products, easy to use, and with a broad field of action led, at first, to an irrational use, too often repeated as a kind of all-risk protection, even in the absence of pests. This systematic or anarchic implementation of protective treatment quickly led to pollution and environmental problems. These pesticides have contributed to early outbreaks of pests resulting from the destruction of native entomophagous organisms, the emergence of pest populations resistant to various groups of insecticides, and pesticide residue accumulation in food products. New approaches have been required and adopted as an alternative to reduce pesticide impact on the environment. These are Integrated Pest Management (IPM) strategies. According to FAO, Integrated Pest Management represents a crop management system, in the environmental context and changing pest populations, using all available control techniques, in the most consistent manner possible, to maintain pest levels below the threshold of economic harm. This management must ensure that there is no complete dependence on a single control method and take into account the economic, social, and environmental consequences of control strategies.

This book describes novel methods adopted in pest management for cereal crops and fruit trees. Each chapter has been written by experts in their respective areas and provides a rigorous review and outline of current trends and future needs, to expedite progress in the field. We have structured the nine chapters of *Pests Control and Acarology* into three sections. In the first section, there is the definition and concept of Integrated Pest Management (Chapter 1) and biological control (Chapter 2). The second section includes two chapters: the first one presents the Trap Barrier System (TBS) as a new tool for rodent pest management in irrigated rice, while the second one discusses the stem borers of cereal crops in Africa and their management. The third section presents strategies for controlling mites.

I would like to thank all contributors for the time and effort that they devoted to prepare their chapters. It is their participation that makes our efforts to organize such a book possible. In addition, special thanks to IntechOpen's staff and editorial board.

Dalila Haouas
University of Jendouba,
Tunisia

Levente Hufnagel
Szent István University,
Hungary

Section 1

Definition and Concepts

Chapter 1

Integrated Insect Pest Management

Hamadttu Abdel Farag El-Shafie

Abstract

Insect pests cause substantial losses to food and fiber crops worldwide. Additionally, they vector human and domestic animal diseases. The dependence on pesticides as a sole method of control has resulted in the development of insect resistance and negative effects on human health, natural enemies, and the environment. The concept of integrated pest management (IPM) originated almost 60 years ago in response to these negative impacts of pesticides. Currently, IPM is a robust paradigm of pest control around the globe. This chapter reviews the history of IPM, its main principles, decision-making rules, the components, and main tactical methods used. Innovative tactical methods such as sterile insect technique (SIT), incompatible insect technique (IIT), and push-pull strategy are discussed. Moreover, challenges of implementation and future prospects of IPM are highlighted.

Keywords: insect pest, integrated pest management, economic threshold, economic injury level, decision rules, ecosystem

1. Introduction

Insects appeared on earth about 390 million years and have diversified into several million species that have adapted to almost all available ecosystems. This large diversity has allowed them to compete with humans effectively since the introduction of agriculture over the last ten millennia [1]. Based on new methods of estimation, there are about 5.5 million species of insects on the earth planet with 1 million identified species, which represent only about 20% of the total estimated number. Previously, the global number of insects was estimated to be 30 million based on host specificity; however, this number seems to be not true [2]. Insects are by far the most successful group of animal on earth and are thus essential component of the ecosystem both economically and ecologically as they make up more than 75% of the world's animal species. Entomology has tremendously developed in recent years and contributed much in the development of other fundamental biological sciences. Today, many insect species are being used as model organisms to study the genomic and proteomic of many organisms. Invasive insect species such as the red palm weevil (*Rhynchophorus ferrugineus*), the fall army worm (*Spodoptera frugiperda*), the spotted drosophila (*Drosophila suzukii*), and the brown marmorated stink bug (*Halyomorpha halys*) are expanding their geographical range and, thus, threatening agricultural crops at a global level [3–6]. According to their mode of nutrition, insects are classified into different categories including herbivores, predators, fungivores, and scavengers. Insects together with weeds and diseases destroy about

40% of the world food production during preharvest phase while approximately 20% is lost during storage [7]. The estimated global losses due to insect pests are 500 billion US$ and by adopting good pest management practices, the losses can be reduced by 42.6% [8].

Well before 2500 B.C., the Sumerians were using sulfur compounds to control insects and mites. By 1200 B.C., the Chinese developed plant-derived insecticides or what is called botanicals today for seed treatment and fumigation uses. They also used chalk and wood ash for prevention and control of both household and stored product pests. In late 1940s, DDT was discovered as a powerful insecticide announcing a new era of pest control [9]. The heavy use of chemical pesticides caused serious environmental problems without achieving final solutions to insect pest problems. These drawbacks of the unwise use of pesticides inspired entomologist to think of integrated pest management (IPM) in 1959 as a new paradigm of insect control [10].

2. Origin and history of IPM

The concept of IPM emerged about 60 years ago when entomologists from California, USA observed that the sole use of chemical pesticides could not be the solution to insect pests' problem. Insect resistance to organosynthetic insecticides, resurgence of primary pests, upsurges of secondary pests, and environmental pollution initiated the notion of IPM [11]. It has been emphasized that chemical control should be employed to reduce a pest population only when natural controls are inadequate. Intervention to control pest should also be made when populations rise to levels that cause economic damage. Additionally, the cost of control must cover the amount lost due to the pest damage and negative effect on the ecosystem, due to the application of pesticide, and should be to the minimum [12]. The IPM concept has three basic elements:

1. maintaining insect populations below levels that cause economic damage;

2. using multiple tactics, in an integrated fashion, to manage insect populations; and

3. conserving environment quality.

As shown in **Table 1**, the publication of the book "silent spring" is considered one of the most important events that hastened the perception of IPM as a new paradigm of pest control. The adoption and support given to IPM by the FAO in 1967 is a major factor behind the development of IPM. Additionally, the establishment

Date	Event	Reference
Late 1940s	The concept of supervised control	[13]
1959	The concept of integrated control	[10]
1961	The Australian ecologists proposed the term "*Pest management*"	[14]
1962	Publication of the book "*Silent Spring*"	[15]
1966	The term "Pest Management" received recognition in USA	[16]
1967	The term "Integrated Pest Management" was used by Smith and Van den Bosch	[17]
1967	FAO panel of experts accepted the term "Integrated Pest Control" as a synonym for Integrated Pest Management	[18]

Date	Event	Reference
1969	The US National Academy of Science formally accepted the term Integrated Pest Management	[16]
1972	Integrated pest management and its acronym IPM were incorporated into English literature and accepted by the scientific community	[11]
1972	The report *Integrated Pest Management* prepared by the Council on Environmental Quality was published	[19]
1988	Major IP success in rice systems in Indonesia	[8]
1989	Farmer Field School (FFS) became a preferred extension methodology for IPM	[20]
1992	IPM was recommended for pest management under Agenda 21 of the United Nations Conference on Environment and Development	[21]
1993–2000	IPM initiative of the Clinton Administration	[22]
2011	IPM programs are operational in more than 60 developing and developed countries	[23]
2011	IPM adopted as a policy within sustainable use of pesticides Directive 91/414/EEC in the form of regulation (EC Regulation 1107/2009), which came into force in June 2011	[12]
2014	The EU Frame work Directive on sustainable use of pesticides (Directive 2009/128/EC) requires that all EU Member States develop a National Action Plan, which ensures that asset of eight general principles of IPM are implemented by all professional pesticide users starting from January 1, 2014	[24]

Table 1.
History and chronological development of IPM.

of Farmers Fields Schools (FFS) in 1989 for rice field in Asia, as extension methods, hastened the adoption and application of IPM at farmer level. Recently, the European Union has adopted IPM as a policy for management of insect pests.

The integrated pest management is now the ideal system for protection of agricultural crops, domestic animals, stored products, public health, and the structure of human dwellings against the attack of arthropod pests, plant and animal diseases, and weeds [1, 11, 25].

3. IPM definitions

Between 1959 and 2000, 67 definitions of IPM appeared in the literature, most of them included using natural or ecologically sound principles or techniques, preventing pests from reaching the economically damaging levels, and using multiple tactics such as cultural, biological, and chemical. The expression economics, environment, pest populations, and pest control appeared in these definitions of IPM with frequencies of 53.8, 48.1, 40.4, and 38.3%, respectively [25]. All IPM definitions include the following: (i) the appropriate selection of pest control methods and decision rules for selection, (ii) economic benefits to growers and society, (iii) the benefits to the environment, and (iv) considering the impact of pest complex [10, 11, 18, 24].

4. Objectives of IPM

IPM has three main objectives: first, maintaining a balanced sustainable ecosystem and a healthy environment by reducing the use of pesticides and their negative impacts; second, saving money by reducing chemical pesticides inputs, crop losses due to insect damage and eventually by reducing the pest management cost; and third, protecting human and animal health by providing food and feed that is free of pesticide residues [26].

5. General principles of IPM

According to the EU Framework Directive 2009/128/EC, there are eight principles of IPM that should be strictly followed by all members of the European Union starting from January 2014 [15]. Barzman et al. [27] described these principles as follows:

5.1 Prevention and suppression

The first line of defense in IPM is to prevent and suppress insect pest population through nonchemical methods such as cultural practices, use of resistant varieties, proper irrigation and fertilization, and natural enemies.

5.2 Monitoring

Continuous surveillance and monitoring of insect pests population is essential for assessment of damage and for determining the needs for actions to be taken.

5.3 Decision-making

Management decisions should be based on monitoring and population levels of insect pests, as well as reliable thresholds.

5.4 Nonchemical methods

Sustainable biological, physical, and other nonchemical methods must be preferred to chemical methods if they provide satisfactory pest control.

5.5 Pesticide selection

Selective pesticides, which have minor negative impacts on human health and beneficial insects, shall be used only when needed.

5.6 Reduced pesticide use

Pesticide use should be kept to the minimum through reduction of doses and application frequency without encouraging resistance development in pest populations.

5.7 Antiresistance strategies

Pesticide resistance in insect should be managed carefully using strategies such as application of pesticides with different modes of action.

5.8 Evaluation

The success of control tactics must be measured using indicators based on monitoring of harmful organisms, beneficials, pesticide use, and impact on the environment.

6. Important terminologies in IPM

Pest: any organism that causes damage or inconvenience to human or his possessions [28].

Natural enemy of a pest: a natural enemy of a pest can be predator, parasitoid, nematode, and pathogens (bacteria, viruses, fungi) [29].

Population: a group of individuals of the same species in a given area that provides the ecological requirements of the species [10].

Population regulation: the return of a population to an equilibrium density, following departure from that density, because of density-dependent processes [28].

Insect pest complex: the number of insect species associated with a particular crop [28].

Major insect pest (key pest): species of insects, which has a high reproductive potential and is capable of causing economic damage on their host [28].

Minor insect pest: insect species that is not capable of causing damage of economic importance.

Secondary or sporadic insect pest: insect species with population level that occasionally grows beyond its economic injury threshold [28].

Economic damage (ED): occurs when the cost of preventable crop damage exceeds the cost of control. For example, if the wheat is worth $ 10 a bushel and insecticide cost $ 15 an acre, then economic damage occurs when insect damage causes a yield loss of 1.5 or more bushels an acre (Ed = cost of treatment/crop value = $ 15/$ 10/bushel = 1.5 bushel).

Economic-injury level (EIL): the lowest population density that will cause economic damage. Economic damage is the amount of injury, which will justify the cost of artificial control measures; consequently, the economic-injury level may vary from area to area, season to season, or with man's changing scale of economic values.

Economic threshold (ET): the density at which control measures should be determined to prevent an increasing pest population from reaching the economic-injury level. The economic threshold is lower than the economic injury level to permit sufficient time for the initiation of control measures and for these measures to take effect before the population reaches the economic-injury level [10] (**Figure 1**).

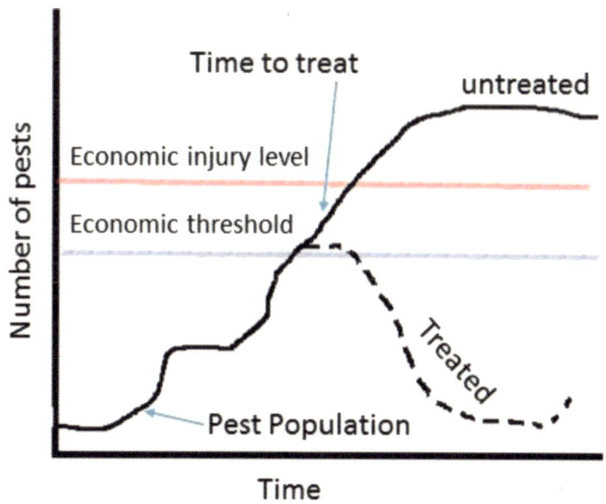

Figure 1.
The economic threshold and the economic injury level.

7. Decision rules in IPM

Identification of pest is essential to gather information about its biology, ecology, and behavior and monitoring population levels. Monitoring includes various activities and procedures that detect and document the presence, growth, and population development or populations levels of an organism. Monitoring is the key to a successful IPM program. Adequate monitoring tools should include trappings using pheromones and light traps, observations in the field as well as scientifically sound warning, forecasting, and early diagnosis systems [27]. Advantages of pest monitoring include early warnings, detection of presence and distribution of pests and their natural enemies, study the impact of weather and other environmental factor on pest/beneficial populations, provision of historical record of the farm, and evaluation of control programs [30]. Visual counts, sweep nets, drop sheets, and vacuum pumps are also useful tools in sampling of field insects.

El-Shafie and Faleiro [31] gave comprehensive accounts on the use of semiochemicals in monitoring and mass trapping of insects. Operational monitoring program is used in IPM to evaluate field situation and should be simple, quick, cost-effective, and adaptable to farmers [30]. There are four methods of sampling insect in the field: random sampling, point sampling, trap sampling, and sequential sampling. More details on sampling of insect pests are given by Flint and van den Bosch [30].

7.1 Pheromones as a monitoring tool

Pheromone-baited traps are commonly used for population monitoring and for mass trapping because they have the following advantages:

1. pheromones are species specific and are, thus, easy to use by untrained people;
2. they function at both small and large pest populations;
3. suitable for early detection and delimitation of infestation by invasive pests;
4. can be used in estimation of population size and determination of number of generations; and

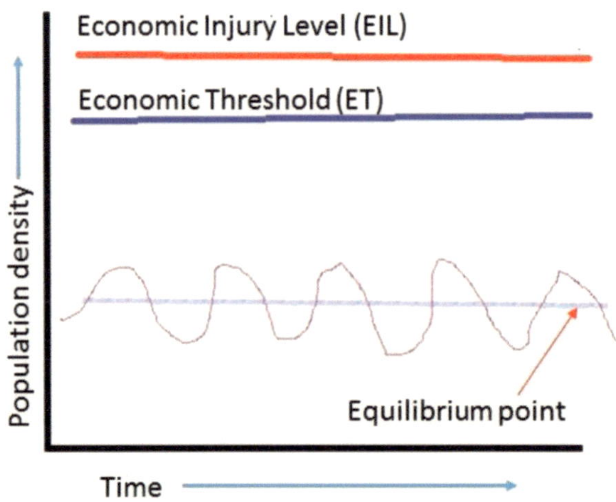

Figure 2.
Equilibrium point is well below the economic injury level. Control action is not needed (modified after Luckmann and Metcalf [33]).

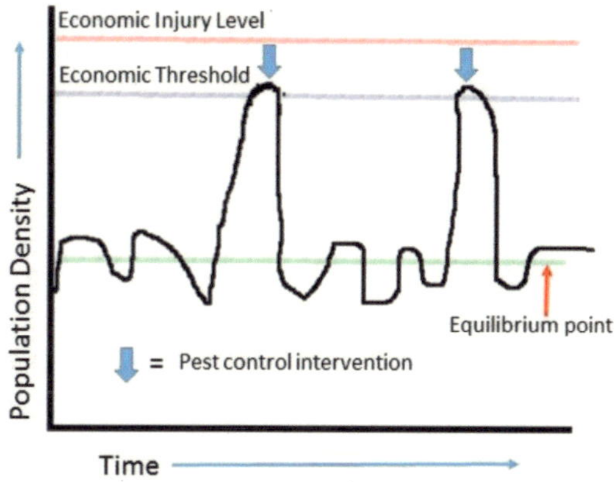

Figure 3.
Equilibrium is below EIL; however, the pest population may reach the ET, and intervention is needed to prevent it from reaching EIT (modified after Luckmann and Metcalf [33]).

5. provide efficient and cheap alternatives to the laborious field scouting.

All of the above make the pheromone-baited traps a useful tool in integrated pest management (IPM) decision-making [32].

In situation where the economic injury level for an insect species is above its equilibrium position (**Figure 2**), the insect is not considered a pest and no decision is needed to control it. However, when the economic injury level is well below the equilibrium position, the insect requires continuous management intervention [33].

Sometimes, the population of the insect pest may reach the economic injury level, even if the equilibrium is well below the economic injury level. In such case, the decision of intervention to control the pest is need to be taken (**Figure 3**). For mathematical calculations of ET and EIL, see Pedigo et al. [34].

8. Components of IPM

The more that you know about a pest, the easier and more successful pest management becomes. Once you have identified a pest, you can access information

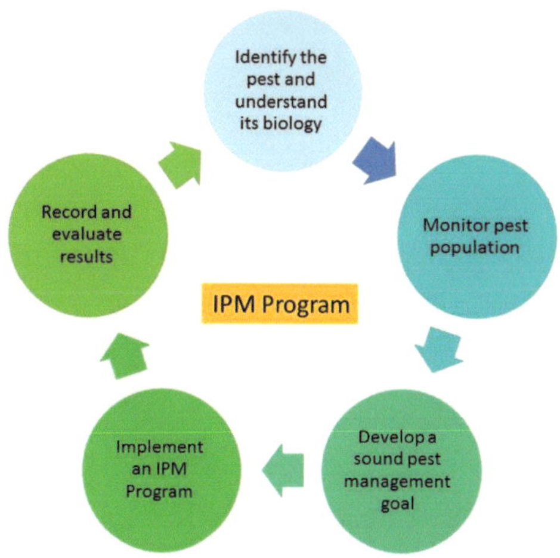

Figure 4.
Components of IPM program.

about its life cycle and behavior, the factors that favor development, and the recommended control procedures. Following identification of an insect pest is monitoring to determine the pest status. If there is a need to control the pest, based on monitoring, then you develop a management program followed by implementation and evaluation as illustrated in **Figure 4**.

9. IPM tactical methods

IPM methods include both chemical and nonchemical means to prevent and control pest populations from reaching economically damaging levels. These prevention and control tactics include biological, mechanical, cultural, physical, genetic, chemical, and regulatory methods. The method to be chosen for IPM depends on many factors, the important of which are nature of target pest, the environment, and economic aspect of the management. Selection of control method should be based on effectiveness and evaluation of any risk that might occur during application of the method.

9.1 Cultural control

Cultural control in cultivated crops include resistant plant varieties, timing of planting and harvesting, irrigation, fertilization, crop rotation, and trap crops. The aim of good cultural practices is to provide congenial environment for the crop while making it unfavorable for pests' development. Thus, cultural control prevents the build-up and outbreaks of pests [28]. Additionally, cultural practices are useful in conservation of beneficial insects, and accordingly, they are essential and effective component of IPM. Tillage practices can destroy pests and their different developmental stages by mechanical injury, desiccation, and exposure to predators and environmental factors [33]. Phytosanitation through collection and removal of crop remains removes many diapausing larvae, eggs, and pathogens. Eradication of infested date palm is a good practice to reduce infestation by the red palm weevil in date palm plantation [3]. Host plant resistance is compatible with other IPM tactics and can provide reasonable degree of protection to plants without causing negative effects on the environment [8].

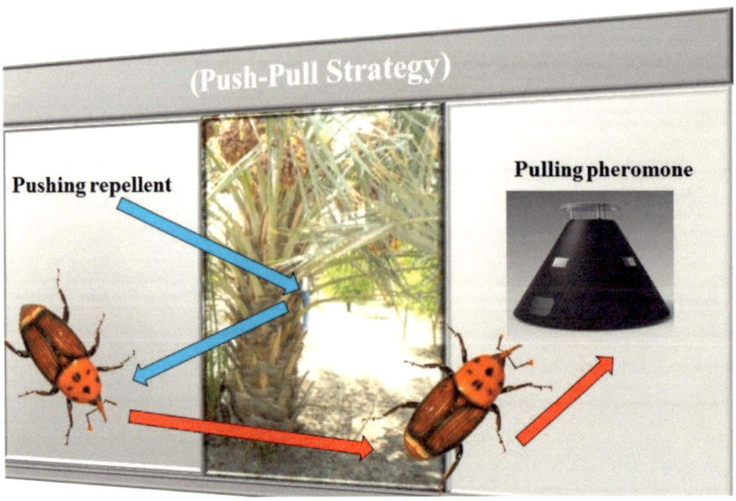

Figure 5.
Push-pull strategy.

9.1.1 Push-pull strategy (PPS)

This strategy is based on intercropping, which fit well under cultural practices section. Simultaneously, it is also based on semiochemicals particularly allomones and kairomones [35]. The pests are repelled or deterred away from a plant (push) through allomones that can be repellents or deterrents and are simultaneously attracted (pull) by kairomones to trap crops where they can be killed or removed [35] (**Figure 5**).

Plants which are effective, so far, in the push-pull tactics include Napier grass (*Pennisetum purpureum*), Sudan grass (*Sorghum vulgare sudanense*), molasses grass (*Melinis minutiflora*), and desmodium (*Desmodium uncinatum* and *Desmodium intortum*). Napier grass and Sudan grass are used for pulling insect pest, whereas molasses grass and desmodium repel or push ovipositing female insects. This strategy has been working in protection of maize and sorghum against damaging stem borers in Africa [36].

9.2 Mechanical and physical control

Mechanical and physical controls prevent pests form accessing their resources by making the environment unsuitable for them. They also negatively affect important biological parameters of pests such as feeding, reproduction, dispersal, and survival. Physical control methods may include heat and steam sterilization of soil, which are commonly used in the management of greenhouse insect pests. Insect pests can be excluded from plants by using screens, barriers, fences, and nets, as well as light trapping (**Figure 6**). Mechanical and physical controls are carried out purposely for pest control, which differentiate them from cultural practices [28].

9.3 Biological control

Biological control is defined as the action of parasites, predators, or pathogens on a host or prey population, which produces a lower general equilibrium, position than would prevail in the absence of these agents [10]. A good biological control agent should be characterized by the following traits: specialization on the host, compatible with other natural enemies, capable of rapid reproduction, adapted to the environment where the host exists, and efficient in finding prey at low densities. There are three major types of augmented biological control: classical, inoculative, and inundative. These are distinguished by the input needed to create a balance between the pest and natural enemy populations. These three categories are defined as follows:

Classical biological control involves introducing natural enemies from a pest's native range into a new area where native natural enemies do not provide control.

Inoculative biological control means releasing natural enemies periodically or seasonally to reestablish a balance that has not been maintained naturally or has been disrupted by other control methods.

Inundative biological control involves the massive production and release of natural enemies to control the pest quickly.

Control of cottony cushion scale (*Icerya purchase*—Maskell) by vedalia beetles (*Rodolia cardinalis*) imported from Australia in 1888 was the first great success and it had greatly benefited the California citrus industry and ignited interest in this practice in the State [37]. Nine species of *Trichogramma* parasitoid are reared in private- or government-owned insectaries around the world and released annually on an estimated 80 million acres of agricultural crops and forests in 30 countries [38]. In Germany and Austria, the control of the Indian meal moth, *Plodia interpunctella* (Huebner), and the Mediterranean flour moth, *Cadra kuehniella* (Zeller), in food

Figure 6.
Solar-powered insect light trap.

processing facilities is achieved by releasing large quantities of *Trichogramma evanescens* Westwood using the inundative release strategy [39].

Manipulations of insect reproductive systems techniques such as sterilized insect technique (SIT) and incompatible insect technique (IIT) provide innovative and environmental-friendly methods for IPM. These techniques are considered as part of the biological control and thus are discussed in this section (**Figure** 7).

The SIT involves the mass release of sterilized males, which mate with wild females. Sterilization of males using ionizing radiation causes dominant lethal mutation in the sperm. The mating of sterile males with wild females results in zero offspring. The sterile insect technique (SIT) has been successfully used for the management of some major insect pests [5]. According to Barnes et al. [40], successful application of SIT depends on the following factors:

1. the target insect pest should be characterized by low population levels;

2. knowledge on the bionomics and genetic of target insect pest;

3. the availability of techniques for mass rearing, releasing, and monitoring of large numbers of viable sterile insects;

4. the release of sterile insects over a wide area to cover the whole population; and

5. the released sterile insects should not be harmful or harmless to humans and the environment.

Another radiation technique is partial male sterility technique (IS), which is used mainly for lepidopterans because full sterilization affects their performance under field conditions. The mating of partially sterilized males with wild females results in sterile male-biased offspring [41].

Wolbachia is an endosymbiont bacterium that is capable of manipulating the reproduction of its host insect. It increases the frequency of *Wolbachia*-infected

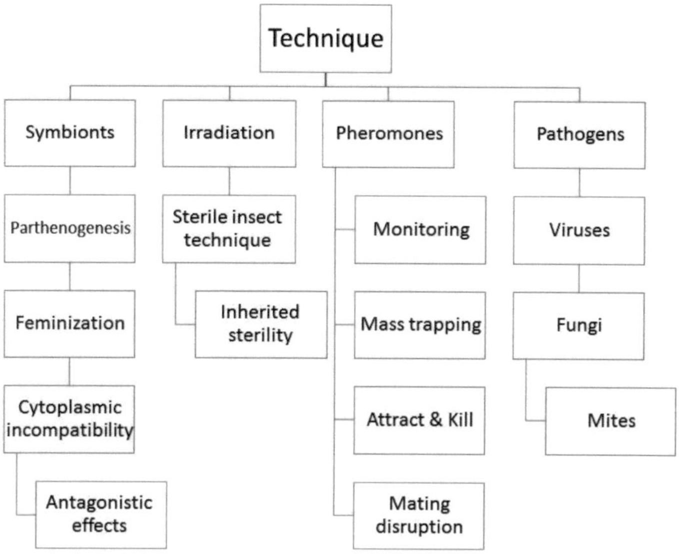

Figure 7.
Techniques of manipulating sexual reproduction of insect pests for their management (modified after Harari et al. [32]).

females in the host populations by causing feminization, parthenogenesis, male killing, and cytoplasmic incompatibility [42]. The cytoplasmic incompatibility (CI) is also called incompatible insect technique (IIT) and has been used against insect pests and disease vectors such as med fly, tsetse fly, and mosquitoes. A strain of *Wolbachia* was taken from *Drosophila melanogaster* and introduced into the mosquito *A. albopictus*, the vector of the dengue virus, in order to control the disease. Consequently, the mosquitoes become unable to transmit the dengue virus [43]. The infected males produce no offspring after mating with local females (CI), followed by a decrease in the local mosquito populations and a relative increase in *Wolbachia*-infected females that do not transmit the virus [44].

Both SIT and IIT can be combined together, and they are compatible with conventional biological control using parasitoid, predators, and pathogens. SIT allows both sexes to be released, while in case of IIT, only males should be released. The release of *Wolbachia*-infected females may result in production of viable offsprings if the released females are compatible with either wild or released males [5].

9.4 Chemical control

Pesticides should only be used when necessary to keep pest populations below that cause economic damage. Selective pesticides, which have the least negative effects on the environment, should be used according to principles 5, 6, and 7 of IPM. Botanicals and microbial (biorational) pesticides should be given priority in selection. The efficacy of these biorational pesticides may be increased when applied together [27]. A variety of selected pesticides must be applied precisely in the field and at right doses to prevent the development of insects' resistance [26].

10. AW-IPM

The integration of a number of different control tactics into IPM systems can be done in ways that greatly facilitate the achievement of the goals either

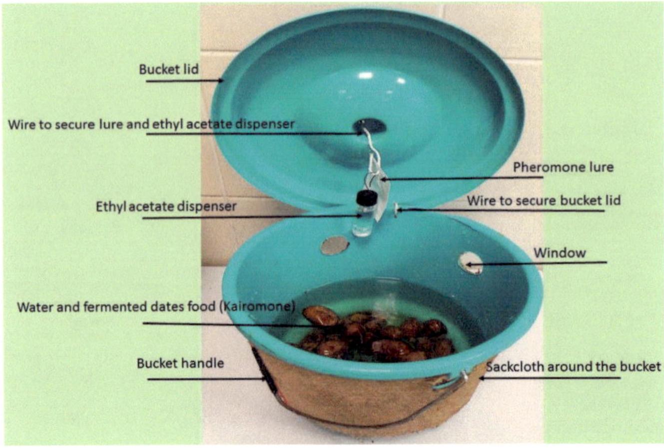

Figure 8.
Pheromone-baited trap for monitoring and mass trapping of red palm weevil.

of field-by-field pest management, or of area-wide (AW) pest management, which is the management of the total pest population within a delimited area [1]. Knipling [45] used simple population models to demonstrate that small insect pest population left without management can compromise the efforts of containment of pest population in a large area. AW-IPM programs should be coordinated by organizations rather than by individual farmers to insure full participation in the program [46]. Pheromone-based control tactics including mass capturing of using pheromone traps (**Figure 8**) proved to be effective against a variety of insect pests in area-wide IPM programs. The pests' behavior and ecology including their natural enemies should be considered when planning future AW-IPM programs [32].

11. Implementation of IPM program

Successful IPM depends mainly on basic research on ecosystem and the understanding of interactions among hosts, pests, and their natural enemies [11]. The following steps should be taken before implementing an IPM program:

- identify the pest;
- specify the goal of the program;
- set up a monitoring program;
- know the pest level that triggers control;
- know what control methods are available; and
- evaluate the benefits and risks of each method.

The socioeconomic factor is important in the implementation of IPM. For example, the decision to include a new variety resistant to insects may also depend on the market value of that variety. A suitable extension methodology such as Farmer Field School (FFS) can help disseminate the IPM among farmers. Preparation of guidelines that include the principles of IPM for different crops is essential during the implementation phase. Moreover, the continuous evaluation of IPM programs provides feedback for future adjustment and improvement [27].

12. Indicators of measuring impact of IPM

It is extremely important to record and evaluate the results of your control efforts. Some control methods, especially nonchemical procedures, are slow to yield measurable results. Other methods may be ineffective or even damaging to the target crop, animal, treated surface, or natural predators and parasites. Pesticide use by volume, pesticide use by treatment frequency index, reduction in use of more toxic pesticides, and environmental impact quotient have been used as IPM impact evaluation indicators [22].

13. Future prospects of IPM

Since 1959, no major departures from the basic notion of IPM have occurred [11]. In the future, major advances in IPM are expected in decision-making techniques as well as tactical options for control methods. Combination of technologies and tools, simulation, modeling, BD, remote sensing data, Geographical Information System (GIS), Automatic Weather Stations (AWS), and internet of things (IoTs) can be used to promote the implementation of IPM. New generation of GPS, sensors-fitted farm equipment, e-tablets, and mobile applications (*Plantix*) could be used for future pests and diseases identification and monitoring [47]. Since implementation of IPM programs depends largely on information, it is anticipated that a giant step being taken in areas such as principles of insect sampling, computer programming and mathematics, understanding of pests biological and ecological aspects, and simulation techniques and modeling [11]. Additionally, meteorological and geostatistical computer models can revolutionize forecasting and monitoring of insect pests, which, eventually improve decision-making for IPM. Novel tactics such as silencing of pest gene or RNA interference (RNAi) and endosymbionts hosted by insect pests could be used as potential new tools for future management of insect pests. Continuous training and education of farmers represent the cornerstone for establishment of solid and effective IPM program in agroecosystems.

14. Conclusions

Due to its importance, the European Union has adopted IPM as a policy for management of insects and other pests. Manipulating reproduction of insect pests with pheromones, irradiations, *Wolbachia*, and pathogens will provide a variety of innovative tactical methods for IPM. Transgenic plants resistant to insect pest are also important tactical methods for future implementation of IPM. The information and communication technology (ICT) and nonprint media such as projectors, tablets, laptops, and mobile cell phones are expected to play a vital role in disseminating IPM knowledge among illiterate farmers, in their languages, in developing countries. In this respect, Julia and Robert [48] started a university-based scientific program called scientific animation without borders (SAWBO) to deliver IPM strategies in the tropics. The SAWBO App helps trainers to use IPM-animated videos, in different local languages, to educate farmers efficiently. The advancement in semiochemical-based tactics could provide great support for area-wide IPM (AW-IPM), which will gain importance in the coming years due to the increasing numbers of invasive insect pest species.

Author details

Hamadttu Abdel Farag El-Shafie
Date Palm Research Center of Excellence, King Faisal University, Kingdom of Saudi Arabia

*Address all correspondence to: elshafie62@yahoo.com

References

[1] Hendrichs J, Kenmore P, Robinson AS, Ureysen MJB. Principles, practice and prospects. In: Vreysen MJB, Robinson AS, Hendrichs J, editors. Area-Wide Control of Insect Pests, from Research to Field Implementation. Dordrech, The Netherlands: Springer; 2007. pp. 3-33

[2] Stork NE. How many species of insects and other terrestrial arthropods are there on earth? Annual Review of Entomology. 2018;**63**(1):31-45. DOI: 10.1146/annurev-ento-020117-043348

[3] Faleiro JR. A review of the issues and management of the red palm weevil *Rhynchophorus ferrugineus* (Coleoptera: Rhynchophoridae) in coconut and date palm during the last one hundred years. International Journal of Tropical Insect Science. 2006;**26**:135-150

[4] Nagoshi RN, Fleischer S, Meagher RL, Hay-Roe M, Khan A, MuruÂa MG, et al. Fall armyworm migration across the Lesser Antilles and the potential for genetic exchanges between north and south American populations. PLoS One. 2017;**12**(2):e0171743. DOI: 10.1371/journal.pone.0171743

[5] Nikolouli K, Colinet H, Renault D, Enriquez T, Mouton L, Gilbert P, et al. Sterile insect technique and *Wolbachia* symbiosis as potential tools for the control of the invasive species *Drosophila suzukii*. Journal of Pesticide Science. 2017;**91**(2):489-503. DOI: 10.1007/s10340-017-0944.y

[6] Nixon LJ, Morrison WR, Rice KB, Brockerhoff EG, Leskey TC, Guzman F, et al. Identification of volatiles released by diapuasing brown marmorated stink bug, *Halyomorpha halys* (Hemiptera: Pentatomidae). PLoS One. 2018;**13**(1):e0191223. DOI: 10.1371/journal.pone.0191223

[7] Pimentel D, Peshin R, editors. Integrated Pest Management, Pesticide Problem. Vol. 3. Dordrecht: Springer Science + Media; 2014. DOI: 10.1007/978-94-007-7796-5

[8] Dhaliwal GS, Koul O, Arora R. Integrated pest management: Retrospect and prospect. In: Koul O, Dhaliwal GS, Cuperus GW, editors. Integrated Pest Management: Potential, Constraints and Challenges. London: CAB International; 2004. pp. 1-20

[9] Flint ML, Van den Bosch R. A history of pest control. In: Flint ML, Van den Bosch R, editors. Introduction to Integrated Pest Management. 1st ed. New York: Plenum press; 1981. pp. 51-81

[10] Stern VM, Smith RF, van den Bosch R, Hagen KS. The integrated control concept. Hilgardia. 1959;**29**:81-101. http://ucanr.edu/repository/fileaccess.cfm?article=152499&p=RNIYON. DOI: 10.3733/hilg.v29n02p081

[11] Kogan M. Integrated pest management: Historical perspective and contemporary developments. Annual Review of Entomology. 1998;**43**:243-270. DOI: 10.1146/annurev.ento.43.1.243

[12] Matthews GA. IPM-a strategy since 1959-mini review. CAB Reviews. 2018;**13**(012):1-2. DOI: 10.1079/PAVSNNR201813012

[13] Smith RF, Smith GL. Supervised control of insects. California Agriculture. 1949;**3**:12

[14] Geier PW, Clark LR. An ecological approach to pest control. In: Proceedings of the Eighth Technical Meeting. Warsaw: International Union of Conservation of Nature and Natural Resources; 1961. pp. 10, 1960-18

[15] Carson R. Silent Spring. New York: Fawcett Crest; 1962. 368 pp. DOI: 10.1097/00005053-196212000-00009

[16] NAS (National Academy of Science). Insect-pest management and control. In: Principles of Plant and Animal Pest Control. Natl. Acad. Sci. Pub. 1695. 508 pp. Vol. 3. 1969. pp. 448-449. DOI: 10.1203/00006450-196909000-00008

[17] Smith RF, van dan Bosch R. Integrated control. In: Kilgore WW, Doutt RL, editors. Pest Control: Biological, Physical, and Selected Chemical Methods. New York: Academic Press; 1967

[18] FAO. Report of the first session of the FAO Panel of Experts on Integrated Pest Control, Rome (Italy); Sept. 18-22, 1967. 19 pp

[19] Council on Environmental Quality. Integrated Pest Management. Washington, DC: Council Environ. Qual; 1972. 41 pp

[20] Peshin R, Bandral RS, Zhang WJ, Wilson L, Dhawan A. Integrated pest management: A global overview of history, programs and adoption. In: Peshin R, Dhawan AK, editors. Integrated Pest Management: Innovation Development Process. The Netherlands: Springer Science + Business Media B.V.; 2009. pp. 1-49. DOI: 10.1007/978-1-4020-8992-3

[21] UNCED. Promoting Sustainable Agriculture and Rural Development. Agenda 21, Chapter 14. United Nations Conference on Environment and Development (UNCED) Switzerland; 1992

[22] Peshin R, Zhang WJ. Integrated pest management and pesticide use. In: Pimentel D, Peshin R, editors. Integrated Pest Management, Pesticide Problem. Vol. 3. Dordrecht: Springer Science + Business Media; 2014. pp. 1-46. DOI: 10.1007/978-94-007-7796-5

[23] FAO. Save and Grow. Food and Agriculture Organization; 2011. The FAO Online Catalogue: http://www.fao.org/docrep/014/i2215e/i2215e.pdf.

[24] Union E. Directive 2009/128/EC of the European parliament and of the council of 21 October 2009 establishing a framework for community action to achieve the sustainable use of pesticides. Official Journal of the European Union. 2009;**52**:71-86. http://eur-lex.europa.eu/legal-content/EN/TXT/?uri=OJ:L:2009:309:TOC

[25] Bajwa WI, Kogan M. Compendium of IPM Definitions (CID) - What Is IPM and how Is it Defined in the Worldwide Literature. Corvallis, OR, USA: IPPC Publication No. 998, Integrated Plant Protection Center (IPPC), Oregon State University; 2002. DOI: 10.1213/00000539-200202000-00035

[26] Dhawan AK, Peshin R. Integrated pest management: Concept, opportunities and challenges. In: Peshin R, Dhawan AK, editors. Integrated Pest Management: Innovation Development Process. The Netherlands: Springer Science + Business Media B. V.; 2009. pp. 51-81. DOI: 10.1007/978-1-4020-8992-3

[27] Barzman M, Bárberi P, Birch ANE, Boonekamp P, Dacchbrodt-Saaydeh S, Graf B, et al. Eight principles of integrated pest management. Agronomy for Sustainable Development. 2015;**35**:1199-1215. DOI: 10.1007/s1359DOI 10.1007/s1359-015-0327-9

[28] Hill DS. Pests of Crops in Warmer Climates and their Control. The Netherlands: Springer Science + Business Media, B.V.; 2008. 704 pp

[29] Flint ML, Van den Bosch R. What is a pest? In: Flint ML, Van den Bosch R, editors. Introduction to Integrated Pest Management. 1st ed. New York: Plenum press; 1981. pp. 31-50

[30] Flint ML, Van den Bosch R. Practical procedures; IPM

monitoring, decision-making, and tools and techniques of the integrated pest manager. In: Flint ML, Van den Bosch R, editors. Introduction to Integrated Pest Management. 1st ed. New York: Plenum Press; 1981. pp. 121-179

[31] El-Shafie HAF, Faleiro JR. Semiochemicals and their potential use in pest management. In: Shield VDC, editor. Biological Control of Pest and Vector Insects. Rijeka, Croatia: InTech; 2017. pp. 3-22

[32] Harari AR, Sharon R, Weintraub PG. Manipulating of insect reproductive systems as a tool in pest control. In: Horowita AR, Ishaaya I, editors. Advances in Insect Control and Resistance Management. Switzerland: Springer International Publishing; 2016. pp. 93-119. DOI: 10.1007/978-3-319-31800-4

[33] Luckmann W, Metcalf RL. The pest management concept. In: Metcalf RL, Luckmann W, editors. Introduction to Insect Pest Management. New York: John Wiley & Sons; 1975. pp. 3-36

[34] Pedigo LP, Hutchins SH, Higley LG. Economic injury levels in theory and practice. Annual Review of Entomology. 1986;**31**:341-368. DOI: 10.1146/annurev.en.31.010186.002013

[35] Cook SM, Khan ZR, Pickett JA. The use of 'push-pull' strategies in integrated pest management. Annual Review of Entomology. 2007;**52**:375-400. DOI: 10.1146/annurev.ento.52.110405.091407

[36] Bhattacharyya M. The push-pull strategy: A new approach to the eco-friendly method of pest management in agriculture. Journal of Entomology and Zoology Studies. 2017;**5**(3):604-607

[37] Perkins JH. Insects, Experts, and the Insecticide Crisis: The Quest for New Pest Management Strategies. New York: Plenum Press; 1982. 304 pp

[38] Li LY. Worldwide use of Trichogramma for biological control on different crops: A survey. In: Wajnberg E, Hassan SA, editors. Biological Control with Eggs Parasitoids. Oxon, U.K.: CAB International; 1994. pp. 37-53. DOI: 10.1104/pp.106.1.37

[39] Schoeller M. Commercial application of parasitoids to control stored-product pests in Germany and Austria. In: Zdarkova E, Lukas J, editors. Cost Action 842 Biological Control of Pests Insects and Mites, with Special Reference to Entomophtorales, Proceedings of the First Meeting of Working Group 4: Bio-Control of Arthropod Pests in the Stored Products, Lisbon 6-7 September. 2001. pp. 29-32

[40] Barnes BN, Hofmeyr JH, Groenewald S, Conlong DE, Wohlfarter M. The sterile insect technique in agricultural crops in South Africa: A metamorphosis.... But will it fly? African Entomology: Journal of the Entomological Society of Southern Africa. 2015;**23**:1-18. DOI: 10.4001/003.023.0103

[41] Gamble JC, Payne T, Small B. Interviews with New Zealand community stakeholders regarding acceptability of current or potential pest eradication technologies. New Zealand Journal of Crop and Horticultural Science. 2010;**38**:57-68. DOI: 10.1080/01140671003767842

[42] Saridaki A, Bourtzis K. *Wolbachia*: More than just a bug in insects' genitals. Current Opinion in Microbiology. 2010;**13**:67-72. DOI: 10.1016/j.mib.2009.11.005

[43] Blagrove MS, Arias-Goeta C, Failloux AB, et al. *Wolbachia* strain wMel induces cytoplasmic incompatibility and blocks dengue transmission in Aedes albopictus. PNAS. 2012;**109**:255-260

[44] Vavre F, Fleury F, Varaldi J. Evidence for female mortality in Wolbachia-mediated cytoplasmic incompatibility in haplodiploid insects: Epidemiological and evolutionary consequences. Evolution. 2000;**54**:191-200. DOI: 10.1111/j.0014-3820.2000.tb00019.x

[45] Knipling EF. Entomology and the management of man's environment. Journal of the Australian Entomological Society. 1972;**11**:153-167

[46] Knipling EF. Regional management of the fall armyworm *Spodoptera frugiperda*- a realistic approach to insect ecology. Florida Entomologist. 1980;**63**:468-480. DOI: 10.2307/3494531

[47] Himesh S, Prakasa Rao EVS, Gouda KC, Ramesh KV, Rakesh V, Mohapatra GN, et al. Digital revolution and Big Data: A new revolution in agriculture. CAB Reviews. 2018;(021):**13**:1-7. DOI: 10.1079/PAVSNNR201813021

[48] Julia BB, Robert PB. Scientific animations without borders (SAWBO): Animating IPM information and education everywhere. Outlooks on Pest Management. 2018;**29**(2):58-61

Chapter 2

Biological Control of Insect Pest

Talha Nazir, Sehroon Khan and Dewen Qiu

Abstract

Among all the crops, the total loss due to the pests varied for each crop likely for wheat 50%, cotton 80%, maize 31%, rice 37%, potatoes 40%, and soybean 26%. Environmental stewardship and food security are the most important factors that involved in agriculture. In many cases by the misuse of insecticide led to population resurgence, pesticide residues, and pest resistance. The microorganisms like virus, fungus, protozoan or bacterium are the active ingredient in this type of pesticides. Safety of food alludes to the conditions and practices that save the quality of food to anticipate tainting and food borne sicknesses. Natural enemies and botanicals play a vital role to control pests with different mechanisms. Microbial, for example, microscopic organisms, growths, and infections are the major biopesticides being concentrated generally to create contrasting options to chemicals. The number and development rate of biopesticides demonstrate an expanding promoting pattern in recent decades. Biopesticides are host particular and biodegradable bringing about slightest persistency of leftover poisonous quality. Biopesticides make key commitments to IPM and can enormously lessen ordinary pesticides. Nowadays, the globe is working on protein-based biopesticide, and it is very effective method to control the insect pest.

Keywords: pests, biopesticide, food security, microorganism, natural enemies, protein

1. Introduction

1.1 Losses due to pests

Among all the crops, the total loss due to the pests varied for each crop likely for wheat 50%, cotton 80%, maize 31%, rice 37%, potatoes 40%, and soybean 26%. Weeds are the major problem in production, and its effects are almost 34% losses. Pathogens and animal pests are also problem in production, but its loss is less as compared to pests, loss due to pathogens is 16% and loss due to animal pests is 18% [1].

1.2 Use of pesticides

Agricultural pests like weeds, insects, pests, and plant pathogen are managed by using pesticide-insecticide. To control the pests, the cost of machinery, fuel, and labor is reduced [2–4]. The advantages of pesticides are production cost is lower, yield is high, and farmer's revenues become high [5]. For crop production, the pesticides are used worldwide and increased about 20 times from 1960 to 2000 [1], in 2002, the usage of pesticide becomes high and it was 1.0 billion tons, and in 2007, it reached to 1.7 billion tons. China is one of the major producer and also the most intensive insecticide-pesticide user in production of crops in the world [6, 7].

In China, the pesticide producers are more than 2800. There are more than 20 large industries which produce the pesticides with a capacity of 5000–10,000 tons per year. Registered active ingredients are more than 600, and up to 2005, the total products (or formulations) were 22,000. Last year, the total amount for production of pesticide was more than 1 million ton. In the sense of active ingredient, 0.28 million tons and almost 1.4 million tons of formulated products are used and every year 20 million hectares. Nearly 30–40% yield loss could be evaded. There are more than 400 million farmers and 200,000 distributions [8].

1.3 Safety of food in field and on agricultural practice

Pesticides are used to increase the crop production in agricultural farms. But the pesticide also has some sewer effects on human health even death because it contains poison. These residues are more effective on children as compared to adults. Different agricultural practices are used to keep the pest population below the economic threshold level. The use of pesticide application is the most adoptable technique used by farmers for the production of agricultural products. But these chemical pesticides caused many serious problems. Due to the indiscriminate use of pesticides for the better production of agricultural products, food residues remain in these products, which cause health problems in human after consumption of these products [9].

2. Threats of pesticides

2.1 Effect on humans and environment

Asthma, learning disabilities, diabetes, birth defects, autism, reproductive dysfunction, Parkinson's and Alzheimer's diseases, and many types of cancer are the most common diseases that affect the health of human in twenty-first century. The chronic diseases and death rate are about 1 million/year people due to the pesticides because they contain poison [10]. The residues of pesticide remain in or on the food after they applied on crops [11]. In many countries, it were seen that the levels of these residues in foods are often stipulated by regulatory bodies. Now, people are going to aware about the residues that occur in the edible food. Several pesticide residues especially exhibit bioaccumulation and derivatives of chlorinated pesticides which could develop harmful effect in human and animal body, also in environment [12]. Pesticides-insecticides can contaminate turf, soil, water, and other vegetation. For reducing or finishing the insects-pests or weeds, we use pesticides-insecticides but it will also be harmful for the other organism like beneficial insects, birds, and also nontarget crop or plants. Pesticides-insecticides contain acute toxic, but herbicides also pose risks to other plants or nontarget organisms [13]. One of the environmental science books is Silent Spring and the author of this book is Rachel Carson, which was published in 1962. In the book, the harmful effects on the environment, predominantly on birds of excessive use of pesticides, Rachel Carson accused the chemical industry for scattering half-truth and public officials of accepting industry claims without any question [14].

3. Types of insecticides

3.1 Transfer processes

Adsorption is the process in which insecticide fix (bind) with soil particle. According to the law, different charges can attract each other, this can be done because insecticides have positive charge and soil particles have negative charge, and moisture helps in absorption. Due to insistence of some insecticides, the insecticides keep on in the field soil for a long time and can be absorbed by plants grown in the field. Leaching is the process in which insecticide can move through soil instead of over the surface. An insecticide is dissolved with the irrigation water and then applied to the field plant. So, solubility is the mail factor. Volatilization is the process in which solid or liquid changes into gas. When volatilized, the insecticide can move away from treated area by the help of air. With the help of vapor pressure, we can determine that the insecticide will volatize or not. If the vapor pressure is high in the air then insecticide will be volatile. Spray drift is the process in which the droplets of spray move away from the application site during application. Runoff is the process in which the insecticide can move with water in the sloping surface. Insecticide may move as it is mixed with water or fasten to the soil particles of destroying soil. It relies on the slope of area; moisture content, rely and texture of soil [15].

4. Degradation or breakdown processes

It is the process in which the insecticide can break down with the help of light, microbes, and chemical reactions. This process may be done in hours or may take days or years; the breaking down of insecticide depends upon the chemical characteristics of the insecticide and environmental factors.

Microbial breakdown is the process in which the insecticide can break down with the help of microorganism's likely bacteria and fungi. When favorable conditions are available like warm temperature, appropriate soil moisture and oxygen, and favorable pH, the microbial breakdown increases. Chemical breakdown is the process in which the insecticide can break down with the help of chemical reactions in the soil. The type and also the rate of chemical reactions that happen are influenced by pH levels, soil temperature, and moisture and fixing of insecticide-pesticide to soil. Photodegradation is the process in which the insecticide can break down with the help of sunlight. Almost all the insecticides are break down in sunlight to some level. The breaking down of insecticide depends upon some factors like properties of the insecticide, intensity of light, and intensity of light [15].

5. Biopesticides: an alternate to traditional chemical pesticides

Environmental stewardship and food security are the most important factors that involved in agriculture. In many cases by the misuse of insecticide led to population resurgence, pesticide residues, and pest resistance. By using less opportunities once a legend idea, biopesticides are widely available, and this idea is working very rapidly. The biopesticide is an alternate way to control pests, and this method is environmentally friendly as well as effective [16].

6. Role of biopesticides in China

In 2008, the entire 97 varieties of biopesticides existed in China and cover the market with biopesticide with 6 billion yuan and it is the 10% of total sale of pesticide. At present, in China, some problems occur in commercializing biopesticides. The agricultural production is low that cannot fulfill the demand of farmers. The production technology of biopesticide is not good enough; less ability of innovation and concentration of industry are also low. They just encouraged the use of biopesticides in some varieties such as fruits, leafy vegetables, and melons. Chinese herbs and tea are also encouraged but on a small scale. The Chinese Government must implement hard policies of agricultural environmental and solid measure to promote biopesticide against pest in the production of agriculture [17].

7. Major classes of biopesticides

In biopesticide, there are certain of pesticide derived from natural materials as bacteria, certain minerals, animals, and plants. There are three types of major classes of biopesticides [18].

7.1 Biochemical pesticides

These are naturally occurring substances that control pests by nontoxic mechanisms. The conventional pesticides kill or disable the pest by contrast. Biochemical pesticides have some substance that interferes in mating, like sex pheromones, also different fragrance of plant extract attracts the pests to trap. Because it is sometimes difficult to determine whether a substance meets the criteria for classification as a biochemical pesticide, EPA has established a special committee to make such decisions [18].

7.2 Microbial pesticides

The microorganisms like virus, fungus, protozoan, or bacterium are the active ingredient in this type of pesticides. Each microorganism have specific active ingredient to control the specific pests, but microbial pesticides can control or kill many kinds of pests which damage the crop production. One fungus can control the weeds and other control or kill the insects-pests [18].

7.3 Plant-incorporated-protectants (PIPs)

These are pesticide substances produced by plants from genetic material which are inserted in plant. Scientists take the gene for BT pesticide protein and insert into plant's own genetic material. After inserting the gene of *Bt* bacterium in the plant, the plant prepared the substance that can destroy or kill the pest. The plant's genetic material and protein, but not plant itself, regulated by EPA [18]. An insect-toxic protein, Bb70p, was purified from *Beauveria bassiana* 70 using ammonium sulfate precipitation, ion exchange chromatography, and gel filtration. The protein caused high mortality by intra-hemocoelic injection into *Galleria mellonella*. Thus, Bb70p appears to be an insect toxin protein, demonstrating novelty. Identification of this insect-toxic protein presents potential to enhance the virulence of *B. bassiana* through genetic manipulation [19].

8. Biopesticide characteristics

Biopesticides have some characteristics:

1. Target range of biopesticides is narrow and used for specific problem.

2. Biopesticides work slowly.

3. Biopesticides suppress the pest population rather than eliminate.

4. Application time of biopesticides is somewhat critical.

5. There is limited field persistence and shelf life.

6. Biopesticides often used as amount of IPM programs.

7. They are much more suitable for human as well as for environment than conventional pesticides.

8. No residue problems commonly present by biopesticide.

9. Biopesticide application on food safety

Food safety requires wide application range of biopesticides which includes:

If one organism is feed on other organism, this is called its natural enemy. The beneficial insects are those insects that feed on pests. In order arthropods, for example, mites and spiders are beneficial [20]. Microbial pesticide exists naturally or by genetically changed fungi, protozoans, bacteria, algae, or fungi. This can be used as an alternate method to chemical insecticide, this is very effective. Biological toxin material is derived from microorganism, for example, fungus or bacterium, this is called microbial toxin. These types of microorganism may cause death or rupture the gut of the pest because these entomopathogens are highly toxic. Studies proved that pathogen develops insecticidal toxin that are very much important in pathogenesis [21]. Antibiotics are the substance that stops the growth or kills microorganism, including both fungi and bacteria. In "bactericidal", the antibiotics kill the bacteria, whereas in "bacteriostatic", the antibiotics stop the growth bacteria. It is the success of biotechnology that develops the transgenic crops which are resistance against the major pests and also commercialize the transgenic crops. In first generation, the products include plants with just single insecticidal *Bt* genes, which shows resistance against the major pests of cotton and corn [22]. In the search for alternative solutions to crop protection problems, the interest in plants and their chemobiodiversity as a source of bioactive substances has increased. Plants are capable of synthesizing an overwhelming variety of small organic molecules called secondary metabolites, usually with very complex and unique carbon skeleton structures [23]. For many years for the protection of crops, these substances are used for the benefit of mankind [24]. Plants release some chemicals in environment and when they are used as smother crops, green manures, intercrop, cover crop or mulch, or in rotational sequences grown than it can invade insect-pests and pathogenic diseases and shows the progress in the yield of crops [25]. Botanicals include crude or semi refined extracts and isolated or purified compounds from various plant species and commercial products [23].

10. Microbial pesticides

Entomopathogenic fungi are considered as essential natural regulators in population of different insects. These fungi have potential as mycoinsecticide agents alongside different agricultural insect pests. The mode of actions of these fungi mainly cause infection in body of their host through cuticle penetration, acquisition entrance to the host hemolymph, production of toxins, and grow by consuming nutrients which are existing in the hemocoel to avoid immune responses in insects [26]. A total of 1600 different types of viruses are involved in the infection of 1100 different species of insects and mites. Baculovirus is considered as a distinct group of viruses. About 100 species of insects are susceptible to this group of viruses. The main feature of this virus group is host specification [27]. Bacterial biopesticides are considered as the most common and inexpensive method of microbial pesticides. Different species of butterflies and moths as well as species of flies, beetles, and mosquitoes are generally controlled by these biopesticides. Bacteria which are used as bacterial pathogens to manage insect pest populations are mainly spore-forming and rod-shaped bacteria in the genus *Bacillus* [28]. The EPNsn introduces symbiotic, pathogenic bacteria of the genera *Xenorhabdus* (in the Steinernematidae) or *Photorhabdus* (Heterorhabditidae) into the hemocoel of their hosts following penetration. Subsequent multiplication of the bacteria leads to host death, which can occur within as little as 48 hours of infection. Nematodes infected insects usually have abnormal behavior than uninfected individuals. The nematode can kill its host without its associated bacterium but is unable to reproduce without it. They have also been used with commercial success against citrus root weevil, against turf pests and on mushroom crops (**Figure 1**) [29].

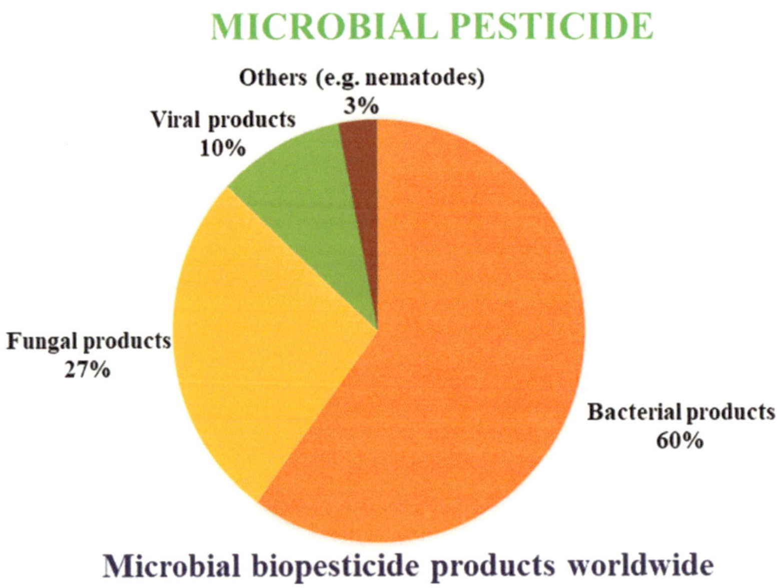

Figure 1.
Microbial biopesticide product worldwide active ingredient: microorganism (bacteria, fungi, virus, nematodes, etc.) (Source: Division of Agricultural Chemicals, 2011).

11. Botanicals

An extensive variety of secondary metabolites were produced by plants that prevent herbivores from feeding on them. These chemicals comprise phenolic compounds (e.g., tannins), glucosinolates (e.g., mustard oil), glucosinolates (e.g., mustard oil), and terpenoids (volatile oils, e.g., limonene). As a plant protection products about 50 different botanical active substances are registered in USA, nevertheless just 11 are registered in the EU. Pesticides are derived from plants and generally act in two ways: that is, stomach poison and contact poison. There are about 250,000 plant species that were evaluated from which 2121 are used in the management of pest, 1005 are demonstrated insecticidal activity, 384 are antifeedants, 297 are repellents, 27 are attractants, and 31 have growth inhibiting properties [30].

11.1 Characteristics of major traditional botanicals

There are some traditional botanical insecticides with their source, mode of action, toxicity, and uses [31] (**Table 1**).

Botanical insecticide	Source plant (s)	Mode of action	Toxicity	Uses
Pyrethrins	Flowers of pyrethrum daisy, *Tanacetum (Chrysanthemum) cinerariaefolium*	Interferes with Na and K ion movement in nerve axons	Mammalian oral LD_{50} > 1000, some allergic reactions can occur	On humans and pests to control lice, fleas, ticks. Rapidly breaks down
Rotenone	Roots of *Derris, Lonchocarpus, other tropical legumes*	Disrupts energy metabolism in mitochondria	Oral LD_{50} = 25–3000 Dermal > 1000	In orchards and gardens against several insects mainly beetles. Persists effectively for 4–5 days or more. Use as a fish poison
Sabadilla	Seeds of tropical European *Veratrum album* and lily *Schoenocaulon officinale*	Interferes with K and Na ion movement in nerve axons. Irritates mucous membranes and skin, potent inducer of sneezing	Oral LD_{50} near 4000	In fruits and vegetables against bugs and citrus thrips. Rapidly breaks down
Ryania	Wood stems of *Ryania speciosa*	Activate Ca++ ion release channels and causes paralysis in muscles of insects and vertebrates	Oral LD_{50} near 1000 Dermal near 4000	In fruit crops and fields against thrips and caterpillars. Frequently combine with pyrethrins and rotenone in commercial mixtures for use of garden

Botanical insecticide	Source plant (s)	Mode of action	Toxicity	Uses
Nicotine	Tobacco, *Nicotiana* sp., *Duboisia, Anabasis, Asclepias, Equisetum, Lycopodium*	Mimics acetylcholine and overstimulate receptor cells to cause convulsions and paralysis	Oral LD_{50} = 3–188 Dermal near 50 Very toxic to humans	Mostly in greenhouses and gardens. Nicotine fumigations target aphids, thrips and mites
Neem	Seeds, leaves, bark of chinaberry (*Melia azedarach*) and neem (*Azadirachta indica*)	Biochemical nature of feeding deterrence, repellence, growth regulation effects are not well described	Oral LD_{50} > 13,000	Medicinally use in humans. On landscape plants and many crops mainly against secondary pests and soft bodied. Very short persistence on treated plants
Limonene/Linalool	Citrus oils	Causes spontaneous stimulation of sensory nerves, biochemical mode of action	Limonene oral LD_{50} > 5000 Dermal > 3500	Mostly in pet shampoos, dip and sprays to kill ticks and fleas. Very short persistence on treated plants

Table 1.
Features of foremost traditional botanicals [31].

12. Natural enemies

If one organism is feed on other organism, this is called its natural enemy. The beneficial insects are those insects, which are feed on pests. In order arthropods, for example, mites and spiders are beneficial. Beneficial arthropods are of different types, parasitoids and predators. Predators like spiders and ladybug; they attach on many kinds of insects and will eat many preys in their life cycle. Parasitoids are flies or wasps that lay eggs inside or on the body of other arthropods, also called parasite. When the egg hatches inside or on the body of other arthropods, the immature parasitoid comes out and feeds on the victim, this is called host, finally killing it. Developing parasitoid, in his life cycle kills only one host. Diseases also affect the insects. Entomopathogens or insect diseases are microorganisms that attack insects and contain nematodes, viruses, fungi, and bacteria. There is some exception that warm-blooded animals cannot affect by the disease attacked by arthropods. To control weeds, sometime plant disease agents and insects are used [20].

There are few natural enemies that are very important from many beneficial insects.

Lacewings found the colonies of aphid to consume just like lady beetles. They feed on scales, mites, mealy bugs, insect eggs, and also on aphids. Stink bugs are called very severe pests and these bugs are useful predators because the eating behavior vary among different species. Searching behavior of lady beetles is frantic; they always bite quickly to eat that helps to understand lady beetles. They feed on many small insects, and no matter that prey is on which stage the aim is just killed. They mostly eat aphid due to his small size but many lady beetles also eat beetle grubs, small caterpillars, whiteflies, mealybugs, scales, mites, and all types of insect egg. From some mites, the spider mites are serious pest of plants but some are beneficial. From all the beneficial mites, the phytoseiid mites are mainly important, the reason is that it is the predator's plant feeding mites and also small organism like thrips or eggs of insect [20].

13. Antibiotics of agricultural

Antibiotics are the active biomolecules obtained from different microorganisms (bacteria, fungi, and actinomycetes). Few antibiotics are derived from plant origin also. Antibiotics kill the pathogens, disease causing harmful microorganisms by interfering with their molecular process like transcription translation, etc., or also by inhibiting the process which are very essential for the survival of pathogens like cell wall synthesis, etc. Antibiotics are further categorized into narrow range and broad spectrum antibiotics. Antibiotics are given orally as well as injected intravenous.

Different antibiotics are used in agriculture. Names of these antibiotics are as follows:

- Validamycin
- Avermectin
- Polyoxin
- Zhongshengmycin
- Wuyimycin
- Agricultural antibiotic 120

14. Insect sex pheromone

The sex pheromones are released by female to attract the male for mating, the female do this with for sexual reproduction. Sex pheromones are for breeding and for attraction of opposite sex and transferring information on their species, sex, age, and genotype after released by male. Volatile pheromones are called as defensive pheromones or sex pheromones, and they have a particular smell and focused on the sensitivity of alarm [32].

14.1 Application of insect sex pheromone

Pheromones are used in the following manners in pest management practical

- Pheromones are used for detecting by surveying and monitoring
- By using pheromones, it is easy to know the resistance of insecticide
- Pheromones are used to attract the pests to that area which is treated with poison or insecticide
- Population suppression by mass trapping
- Pheromones are also used to attract the pest to sterile them

15. Difference between chemical and biological pesticide

Chemical pesticide	Biological pesticide
Nontarget species are also harmed due to chemical pesticides	These do not harm nontarget species
These cause pollution, sometime serious	They do not pollute the environment
Harmful pesticide residues may often remain in food, fodder, and fibers	These have no harmful residues remaining in food, fodder, and fibers
These are relatively costlier	These are relatively cheaper
Insects may become resistant to pesticides, e.g., *Heliothis* has become resistant to most insecticides	Insects are expected not to develop resistance to biopesticides
They are nonspecific to target, so accurate identification of the pest is not necessary	Since they are highly specific, so correct identification of the pest is essential
High specificity often not required to make the use of pesticides	High specificity may often make the use of two or more biopesticides necessary
	Performance may be variable due to the influence of biotic and abiotic factors of the environment

16. Role of biopesticides in organic agriculture

Organic farming is a system used to produce agricultural products like food and fiber. The main aim of organic farming is that to develop biological diversity in the field to disturb the habitat for the organisms of pests, and the purposeful maintenance and replenishment of soil fertility. Biopesticides are used in organic agriculture for minimizing the pest population. To control harmful organisms or pests, different parts of plants are used like chili and garlic due to their strong smell against insects. Against mosquito sweet, basil is used as a repellent and so on. Biopesticides have bright future in the case of organic agriculture [33].

17. Future of biological control

In an environment, natural enemies are responsible for the regulation of 98% pest. From these 98%, only 5% pests have been controlled by the use of

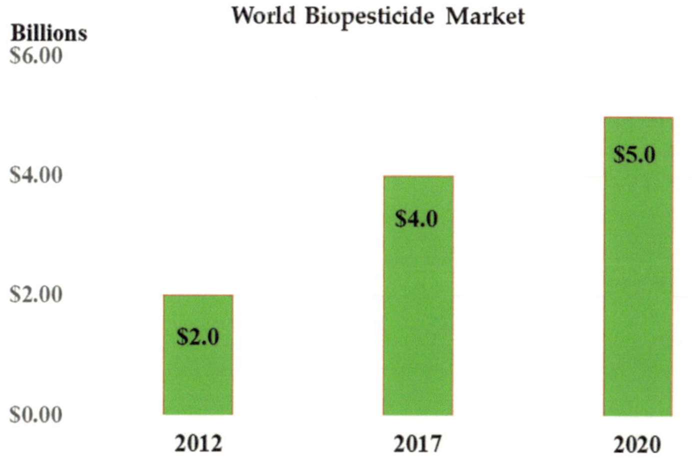

Figure 2.
World biopesticide market, fruit and vegetable crop represent 80% of usage (Source: Piper Jaffray Agricultural Note, August 2013).

entomopathogenic fungus. Reduction of insecticide use and exploitation of bio-control agent can help in the population regulation of different pest and biocontrol agents. The importation of new predators and parasitoids in an environment is also a good source of biological control. The use of some new trends like strains, biotypes, and hybrids of parasites, fungal biopesticide, and use of different viruses also need attentions to be explored properly and can be used for the management of pests. The use of biocontrol agent has numerous advantages, but it is important to modify the other methods of control like the use of pesticides. The pesticides must not be very toxic to biocontrol agents, for example, endosulfan is less harmful to many natural enemies. Biocontrol helps in upholding 'Balance of Nature' as it is the phase of natural control [34].

Biopesticide, from a long time attracting the world consideration as safe and sound plan than chemical control of pest with less hazard to environment and the human. The collaboration between private and public sectors is compulsory to assist the growth, manufacturing, and sale of this environmentally friendly alternative. With research, new substances will be formulated and the delivery will also help in commercialization and usage of biopesticide. In developing countries, one of the most important factors is the availability and quality of the given product must be in low price. New products could help as promising choice against pest; further more research in field is necessary to confirm the efficacy on target pest in many cropping systems [35] (**Figure 2**).

18. Conclusion

Biocontrol is environmentally friendly and active means of decreasing or mitigating pests and pest effects through the use of natural enemies. The goal of b*iocontrol* is to promote the technology and science. Biological control is a technique of controlling pests, that is, mites, insects, weeds, and plant diseases by using other microorganisms.

Biopesticide is used for the modification of development of insect and behavior exerts unique approach for management of insect population. The application of biopesticide is based on the principle that is to provide safety to the human and environment. Wide research is going on, and it is required much more in future to achieve the improvement. The future of biopesticide would fully depend on adoption of application of biopesticide. Versatile use of biopesticide must meet the aims. They must able to control/suppress/kill the harmful insects pests and also prevent them, relatively in proper manner to conventional methods. Researchers, producers, and farmers should widely explore the use and find the safe environment and also maximum production of crops.

Acknowledgements

This work was supported by the National Key Research and Development Program of China (2017YFD0200900).

Conflict of interest

Authors have declared no conflict of interest.

Author details

Talha Nazir[1], Sehroon Khan[2] and Dewen Qiu[1]*

1 Key Laboratory of Integrated Pest Management in Crops, Institute of Plant Protection, Chinese Academy of Agricultural Sciences (CAAS), Beijing, P.R. China

2 Key Laboratories of Economic Plants and Biotechnology, Kunming Institute of Botany Chinese Academy of Sciences, Kunming, Yunnan, P.R. China

*Address all correspondence to: qiudewen@caas.cn

References

[1] Oerke E-C. Crop losses to pests. The Journal of Agricultural Science. 2006;**144**(1):31-43

[2] Osteen CD, Szmedra PI. Agricultural Pesticide Use Trends and Policy Issues. Agric Econ Rep. US Department of Agriculture, Economic Research Service; 1989

[3] Fernandez-Cornejo J, Jans S, Smith M. Issues in the economics of pesticide use in agriculture: A review of the empirical evidence. Review of Agricultural Economics. 1998;**20**(2):462-488

[4] Gardner JG, Nehring RF, Nelson CH. Genetically modified crops and household labor savings in US crop production. AgBioforum. 2009;**12**:303-312

[5] Fernandez-Cornejo J, Nehring R, Osteen C, Wechsler S, Martin A, Vialou A. Pesticide Use in US Agriculture: 21 Selected Crops. EIB-124, U.S. Department of Agriculture, Economic Research Service; 1960-2008. May 2014

[6] Peshin R, Bandral RS, Zhang W, Wilson L, Dhawan AK. Integrated pest management: A global overview of history, programs and adoption. In: Integrated Pest Management: Innovation-Development Process [Internet]. Springer; 2009. pp. 1-49. Available from: http://link.springer.com/10.1007/978-1-4020-8992-3_1

[7] Zhang W, Jiang F, Ou J. Global pesticide consumption and pollution: with China as a focus. Proceedings of the International Academy of Ecology and Environmental Sciences (IAEES); 2011;**1**(2):125

[8] Anastassiades M, Tasdelen B, Scherbaum E, Stajnbaher D, Ohkawa H, Miyagawa H, et al. Pesticide Chemistry: Crop Protection. In: Public Health, Environmental Safety. Weinheim: Wiley-VCH Verlag GmbH Co. KGaA; 2007. p. 439-458

[9] Abrol DP, Shankar U. Pesticides, food safety and integrated pest management. In: David Pimentel RP, editor. Integrated Pest Management: Pesticide Problems. 1st ed. Vol. 3. Springer; 2014. pp. 167-199

[10] Environment K. Killer environment. Environmental Health Perspectives. 1999;**107**(2):1-2

[11] McNaught AD, Wilkinson A. Compendium of Chemical terminology. In: Blackwell Scientific Pubblications. Oxford: Blackwell Science Oxford; 1997

[12] Crinnion WJ. Chlorinated pesticides: Threats to health and importance of detection. Alternative Medicine Review. 2009;**14**(4):347-360

[13] Aktar W, Sengupta D, Chowdhury A. Impact of pesticides use in agriculture: Their benefits and hazards. Interdisciplinary Toxicology. 2009;**2**(1):1-12

[14] McLaughlin D. Silent Spring Revisited. Front Online Spec Rep Fooling with Nat. 2001

[15] Singh DK. Toxicology: Agriculture and Environment. Pesticide Chemistry and Toxicology. Vol. 1. Bentham Science Publishers; 2012. 150 p

[16] Gupta S, Dikshit AK. Biopesticides: An ecofriendly approach for pest control. Journal of Biopesticides. 2010;**3**(special issue):186

[17] Wu L, Zhu D. Food Safety in China: A Comprehensive Review. CRC Press; 2015. 384 p

[18] EPA U.S. Environmental Protection Agency. What Are Bio Pesticides?

[Internet]. 2016. [cited 2016 May 5]. Available from: https://www.epa.gov/ingredients-used-pesticide-products/what-are-biopesticides#classes

[19] Khan S, Nadir S, Lihua G, Xu J, Holmes KA, Dewen Q. Identification and characterization of an insect toxin protein, Bb70p, from the entomopathogenic fungus, *Beauveria bassiana*, using *Galleria mellonella* as a model system. Journal of Invertebrate Pathology. 2016;**133**:87-94

[20] Smith HA, Capinera JL. Natural Enemies and Biological Control. [Internet]. 2017. pp. 1-6. Available from: http://edis.ifas.ufl.edu/in120

[21] Burges HD. Safety, safety testing and quality control of microbial pesticides. In: Microbial Control of Pests and Plant Diseases 1970-1980. 1981

[22] Ferry N, Edwards MG, Gatehouse J, Capell T, Christou P, Gatehouse AMR. Transgenic plants for insect pest control: A forward looking scientific perspective. Transgenic Research. 2006;**15**(1):13-19

[23] Cavoski I, Caboni P, Miano T. Natural pesticides and future perspectives. In: Pesticides in the Modern World-Pesticides Use and Management. Rijeka: InTech; 2011

[24] Rattan RS, Sharma A. Plant secondary metabolites in the sustainable diamondback moth (*Plutella xylostella*) management. Indian Journal of Fundamental and Applied Life Sciences. 2011;**1**:295-309

[25] Farooq M, Jabran K, Cheema ZA, Wahid A, Siddique KHM. The role of allelopathy in agricultural pest management. Pest Management Science. 2011;**67**(5):493-506

[26] Meadows MP. Bacillus thuringiensis in the environment: Ecology and risk assessment. In: *Bacillus thuringiensis*, An Environmental Biopesticide: Theory and Practice. John Wiley and Sons; 1993. pp. 193-220

[27] Usta C. Microorganisms in biological pest control—A review (bacterial toxin application and effect of environmental factors). In: Current Progress in Biological Research. Rijeka: Intech; 2013

[28] Harwood CR, Wipat A. Sequencing and functional analysis of the genome of *Bacillus subtilis* strain 168. FEBS Letters. 1996;**389**(1):84-87

[29] Georgis R, Koppenhöfer AM, Lacey LA, Bélair G, Duncan LW, Grewal PS, et al. Successes and failures in the use of parasitic nematodes for pest control. Biological Control. 2006;**38**(1):103-123

[30] Purohit SS, Vyas SP. Medicinal Plant Cultivation: A Scientific Approach: Including Processing and Financial Guidelines. India: Agrobios; 2004

[31] Weinzierl RA. Botanical insecticides, soaps, and oils. In: Biological and Biotechnological Control insect pests. Boca Raton, Florida: Lewis Publishers; 2000. pp. 101-121

[32] Regnier FE, Law JH. Insect pheromones. The Journal of Lipid Research. 1968;**9**(5):541-551

[33] Sarkar NC. Role of biopesticides in organic farming. International Journal of Agriculture Environment & Biotechnology. 2009;**2**(1):102-104

[34] Agriinfo. http.pdf [Internet]. Available from: http://agriinfo.in/default.aspx?page=topic&superid=6&topicid=751

[35] Damalas CA. Current Status and Recent Developments in Biopesticide Use. Agriculture [Internet]. 2018. **8**(1):13. Available from: http://www.mdpi.com/2077-0472/8/1/13

Section 2

Pest Management in Cereals

Chapter 3

Trap Barrier System (TBS) as a New Tool for Rodent Pest Management in Irrigated Rice in Africa

Loth S. Mulungu, Bernard M. Mchukya and Laurent L. Mnyone

Abstract

Rodent infestation poses a serious threat to smallholder farmers in both developed and developing countries where a large proportion of potential crop yield is lost. In Tanzania, the average annual yield loss of rice is estimated to be around 5–12%. Management of rodent pests in Africa relies mostly on the use of rodenticides which, however, are often applied only when damage has already occurred rather than routinely. Rodenticides used in this way are rarely economically and ecologically sustainable for managing rodents in irrigated rice. A "community-based Trap-Barrier-System (cTBS)" provides an alternative novel rodent control approach for controlling rodents in rice fields. This is basically a system where rodents are trapped in a rice field that is planted a short period earlier than the surrounding fields and therefore attracting rodents from a much wider area than the field itself. The system has proved very successful in irrigated rice fields in Tanzania, increasing rice yields in the intervention fields by 40.91%. A single cTBS can confer protection in up to 16 ha of irrigated rice field. Therefore, if scaled up and used widely, TBS has a great potential for managing rodent pests and improving yield in irrigated rice fields.

Keywords: trap barrier system, rodent pest, irrigated rice, management, community

1. Introduction

Rice (*Oryza sativa*) is among the three leading food crops of the world, with maize (corn) and wheat being the other two. All the three crops provide around 42% of the world's required caloric intake. In 2009, human consumption was responsible for 78% of the total usage of produced rice [1]. More than 3.5 billion of the world's population, which translates to at least half of the people living in the world, thinks of rice as their staple food.

According to IRRI [1], the top rice producing countries include India (43.2%), China (30.35%), Indonesia (12.16%), Bangladesh (12.00%), Thailand (9.65%), Vietnam (7.66%), Burma (6.8%), Philippines (4.5%), Cambodia (2.9%) and

Pakistan (2.85%). These countries are also among the top rice consumers of the world and combine to account for around 90% of the world's rice consumption.

Rice is also one of the most important cereals grown and used as staple food in many African countries [2]. It is the second most important crop in Africa after maize [3]. Rice is produced under typical monoculture systems [4] that can be subdivided into three agro-ecosystems: rainfed lowland (74%), rainfed upland (20%) and irrigated lowland (6%), and the average production is 2.2 t/ha in Africa and 3.4 t/ha worldwide [4]. Farmers in Africa grow mainly local and traditional varieties, many of which have low yield potential. Most of the rice grown depends on rainfall and many irrigation schemes. However, the yield and performance of wet land rice planted in different countries still exhibit wide variations due to the varying climate, land and soil, water supply, farming practices, socio-economic conditions and other biological agents such as rodents [5].

1.1 Impact of rodents to rice crop

Rat damage to ripening rice crops in Asia, Africa and Latin America can be an extremely serious agricultural problem, although economic losses are often difficult to estimate because of complex patterns of growth and recovery of plants related to the developmental stage when damage occurs [5, 6]. Rats can completely consume fields of growing rice and sometimes prevent planting where crops could otherwise be grown [7]. Rodent outbreaks in rice cropping areas have been reported to cause severe crop damage and food shortages [8] due to effects from sowing to physiological maturity of the crop.

In many countries, farmers consider rodents as an inevitable pest in their fields [9]. Thus, they consider chronic rodent damage as something beyond their control [10]. Rodent pest species cause numerous loses in different seasons and locations [11]. However, in some locations, for example, in Philippines, farmers tend to ignore rodent problems on standing rice when cut tillers are less than 5%. Significant reduction in yield is observed at 25% cut tillers when compared with rice field where rodent control is practiced [12]. The authors reported that farmers tend to seek help or apply control measures when rat damage is higher than 5% or when damage occurs at a critical stage of the crop, that is, at milky to soft dough stage.

Rodents, particularly rats, substantially cause damage to rice fields [12]. They eat rice seeds and seedlings (**Figure 1**), gnaw tillers (**Figure 2**), damage plants and feed

Figure 1.
Rice seedling in nursery damaged by rodent pest (Courtesy by Loth S. Mulungu).

Figure 2.
Rice tillers damaged by rodent pest (Courtesy by Loth S. Mulungu).

on grains [13, 14]. In Tanzania, it has been addressed as the major threat in rice crop production system. Farmers keep on controlling the pest to meet household food demands. Elsewhere, on average across Asia, 5–10% of crop damage has been attributed to rodents [9, 12].

Rodent damage to rice can be measured at several stages of crop growth. The level or severity of damage is not uniform throughout growth stages of the crop; instead, it tends to be more concentrated at some growth stages [15, 16]. At planting, for example, rodents may dig up and eat the planted rice seeds in nurseries or in fields which are directly planted and consequently necessitate repeated late replanting [17] and ultimately result in lower yield [11].

At vegetative stage while paddy is growing, rats cut rice tillers and use for building their nests [18] and eat [19]. Damage can be severe during the dry season and cuts are normally seen at the base [15, 16, 20]. At 45° which make different with other pest [21]. At maturity, rodents attack both milky and mature grains [15, 16, 21]. In Asia, an estimated rodent damage of 5–10% was recorded prior to rice harvest in 1999 [22]. In Tanzania, for example, rodents cause an estimated 10–12% pre-harvest loss of rice annually [23, 24].

In Indonesia, rodent pests, primarily the rice field rat (*Rattus argentiventer*), are the most important pre-harvest pests causing annual losses of rice crops by 17% [25]. In Vietnam, My Phung and Brown [26] reported rodent damage on rice to increase from 2.1 (in the first rice crop, winter–spring) to 3.8% in the second (Summer–autumn) rice crop and reached 6.6% in the third (autumn–winter) rice crop and caused yield loss of 15%. In Western Kenya, Taylor [27] reported rodent-associated losses of maize, wheat and barley to be 20, 34–100 and 34%, respectively, during rodent outbreak periods.

In West Java, monocultures of lowland irrigated rice, cumulative damage to rice during the dry season was 54% at the primordial stage, 32% at the booting stage and 16% at the ripening stage. The rodents cause major impacts in agriculture in most parts of the world by attacking crops at any growth stage. However, according to Mulungu et al. [11], the impact of the rodent damage on final yield depends on the country, season and crop type. For example, in Vietnam, rodent pests have been serious since 1995 and considered top three agricultural problems in pre-harvest of lowland irrigated rice [28]. In Indonesia, a loss of 10–20% for pre-harvest was observed each year [28].

2. Rodent management

2.1 Rodent management options

The history of rodent pest management in Tanzania goes back as early as 1912 when rodent (*M. natalensis*) outbreaks were reported in Rombo district in Kilimanjaro region [29]. Studies on population characteristics of this species showed irregular population explosions and most of the outbreaks occurred during the dry season and last through the planting season of October–February [30].

In the past, most of the control measures used in then were localized [31]. With technological advancement and population growth, several changes took place, and at present, rodent control options can be grouped into two basic approaches: the lethal and non-lethal [31]. Many different methods for controlling rodent pests have been passed down through folklore or have been tested and proven effective in particular situations [32, 33].

2.1.1 Non-lethal or preventive measures

The non-lethal method involves habitat manipulation or cultural practices, exclusion/fencing and use of repellants. Environmental sanitation involving the removal of fallow patches in crop fields is another non-lethal practice used in many places [10]. Thick grass and bushes provide harborage and supplementary food resources to rodents. In Tanzania, the environmental sanitation has been done by farmers through slash and burning fields before sowing and harvesting as a way of displacing rodent population [10]. Deep plowing and regular weeding have been reported to suppress rodent population due to destruction of nests, removal of alternative source of food and harborages [34]. However, sanitation is not significantly effective as most farmers practice it on small plots that are interspersed with patches of fallow and permanent grassland [34].

According to Masol et al. [35], the behavioral defense of pest against contact especially for dietary poisoning influences their feeding and area repellent. For area repellent, Voznessenskaya et al. [36] reported the exposure to predator odor to cause disruption of the estrous cycle. Voznessenskaya et al. [37] reported reduced 26 reproductive outputs as the result of exposure to area repellent, specifically urine products derived from meat diets and urine from rats housed in a crowded condition. Mulungu et al. [33, 38] observed significant reduction in rodent activities following

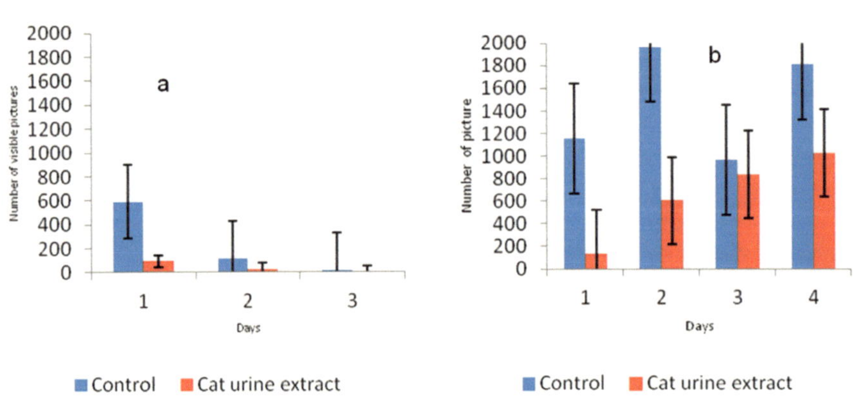

Figure 3.
Effectiveness of cat urine extract of (a) female cat urine extract and (b) male cat urine extract. Source: Mulungu et al. [33].

the application of cat urine. Female cat urine extract repelled significantly more rodents as compared to male cat urine extract (**Figure 3**). The author further reported that the repellent effect was observed from day 1 to 4, but not beyond (**Figure 4**).

In Tanzania, Ngowo et al. [32] evaluated two compounds, that is, thiram and cinnamamide treated in maize seeds as contact repellent, and reported that these two compounds excel over no treated maize seeds in both laboratory against *M. natalensis* and fields against rodent pest species. Mdangi [39] reported that castor oil (*Ricinus communis*) is therefore a promising rodent repellent for small scale maize farmers (**Figure 5**), which protect maize seeds during sowing time.

Another non-lethal method is exclusion or fencing, which is the technology that involves setting of barrier to prevent rodents from reaching the area of concern. It is mostly practiced in smaller areas or in valuable crops like seedbeds and research

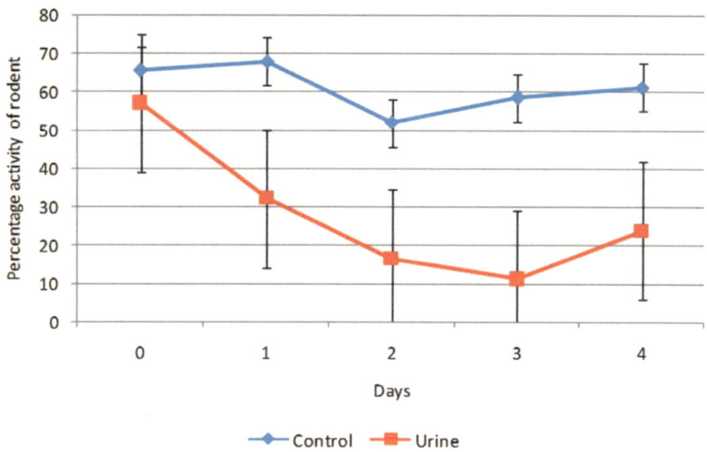

Figure 4.
Percentage rodent activities (±SD) on tracking tiles in rooms treated with either female or male cat urine extract. Source: Mulungu et al. [38].

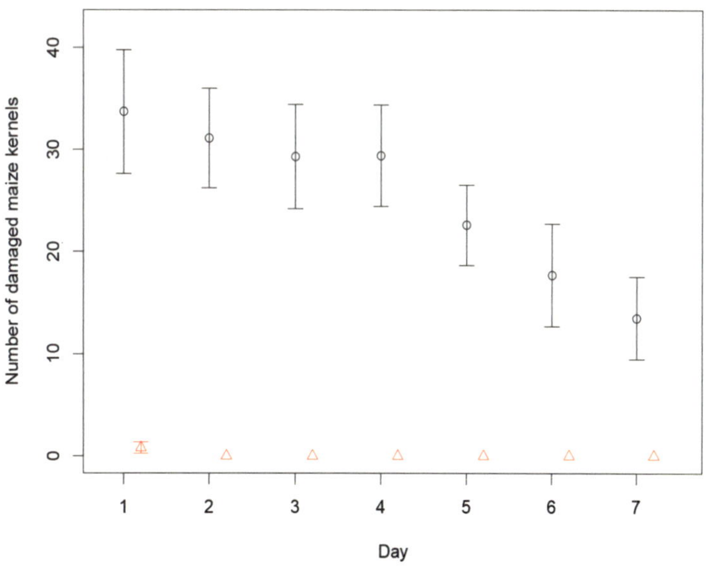

Figure 5.
Mean proportion of damaged treated maize against number of damaged maize seeds in the control group (black) and the treated group (red) at 150 g concentrations of castor oil. Source: Mdangi [39].

plots [40]. Rodent proofing in houses whenever possible is a critical step in controlling rodents. This could be through making it impossible for them to gain entry to the house. It has been reported that fences which relied on the use of barriers that exceeded the physical capability of the rodent pests were reliable [41].

2.1.2 Lethal approaches

The lethal rodent control methods are based on traditional, historical and conventional approaches (e.g. trapping, chemical, toxicants and biological control) [31, 42, 43]. The major methods of achieving satisfactory mortalities are physical killing by trapping and rodenticides [44, 45]. However, killing with rodenticides during rainfall and in irrigation schemes is compromised by water hence loss of effectiveness and increased chances of poisoning non-targeted organisms [45, 46].

Rodenticides and traps are known to provide immediate effects to the problem and are often considered to be the most practical, economical and effective methods of combating rodents. The biological method always requires a period of time before it becomes stable and provides substantial results [47].

The introduction of predators to control pests is an ecologically and conceptually appealing approach for reducing rodent pest populations. Introducing biological agents to control rodents is a promising area for research, but many challenges remain to find a candidate which is sufficiently pathogenic, has a high transmission rate and is target specific [48]. The role of natural predators in controlling rodent pests is an interesting, but frequently misunderstood, concept that is rarely effective in reducing pest populations to tolerable levels [49, 50]. The introduction of barn owls, for example, to Hawaii for rodent control in the 1960s was ineffective. Some studies on barn owl in lowland Southern England revealed that barn owls can adapt and establish to various living conditions in which rodent population exist [51].

In Malaysia, the barn owl was reported to suppress rodents in rice fields resulting in significant lower crop damage [52]. Successful introduction of exotic vertebrate predators into new areas for pest control purposes has never been demonstrated and, in some cases, has resulted in unanticipated, calamitous ecological effects [53]. During the late 1800s, the small Indian mongoose (*Herpestes javanicus*) was introduced into both the West Indies and Hawaii to control rat populations in sugarcane fields [54]. Although this predator survives in some areas on a diet composed mainly of rats [55, 56], the introductions failed to achieve the desired result of reducing rat populations in sugarcane fields.

A variety of traps either commercially available or constructed in homes or villages are used to control rodents; the centuries-long search for "a better mousetrap" has not ended [57, 58]. Trapping is widely used by specialists for surveillance and monitoring of rodent infestations and is, perhaps, the most selective technique to remove individual rodents from problem situations [10].

Although trapping is very labor intensive and requires skill to be used effectively, its relatively low cost compared to other approaches often makes it a primary method of choice for rodent control [59]. Trapping is also utilized where non-target animals are an important concern or where use of toxicants or other more effective methods is prohibited [59]. Trapping generally is not practical for managing large infestations or removing entire populations over extensive areas [60]. However, traps can be used effectively in limited areas or where substantial resources are available and more efficient techniques cannot be used or developed [60]. Farmers, however, try to minimize the crop damage and yield loss caused by rodents by adopting different rodent control methods including poisons (rodenticides), burrow digging to kill rodents, burying buckets full of water, use of live traps and kill traps [23].

Most subsistence farmers rely mostly on the use of rodenticides [61]. Both acute and chronic rodenticides have been used extensively during rodent outbreaks [62]. These chemicals carry significant economic costs and, if used inappropriately, can kill non-target animals (**Figure 6**) and have a negative effect on environment and human health. It can occur when the dead bodies of poisoned rats are eaten by other animals such as birds where the toxin enters the food chain causing death to a variety of other animals including human [10]. Sometimes baiting using acute rodenticides especially zinc phosphate is only used during rodent outbreak [10].

However, rodents are able to multiply fast and re-colonize the farms after rodent control operation [63]. Rodenticides are generally an integral part of successful rodent pest management and, in some tropical habitats, are the only practical method available [64]. Unfortunately, farmers and extension personnel are often confused or uninformed as to how a particular product may be effectively used. In fact, it depends on (i) availability of the required rodenticides, farmers do not access of rodenticides in time when needed, and even if available, they are distributed while damage has already occurred. In some areas, farmers attempt to buy rodenticides from local vendors for control of rodent in their fields themselves. However, most of them report on inefficient control of rodent by the rodenticides they buy, and this is because some vendors sale fake rodenticides prepared from radio dry cell battery and its flour looks like zinc phosphate. Also, improper use of rodenticides and other chemicals for rodent control is a problem whereby farmers lower doses of rodenticides to cover their cultivated areas using few amounts. However, the dose supplied can result to resistance in some rodent species against the commonly and most frequently used chemicals. In some areas, farmers have improper use of chemicals recommended for human being; for example, indocid capsules have been alternatively used by farmers for the control of rodents in fields. (ii) Acceptability of bait formulations to rodents (often influenced by palatability under field conditions). In rodent pest management programs, poison baiting is the most widely used technique throughout the world [65, 66]. Although rodenticides can be incorporated either in bait, dust or water formulations [67], they are generally included in food baits to achieve good control. Much effort has been made to improve the palatability of rodent baits to ensure maximum ingestion by the target rodent pests and thereby improved efficacy. (iii) The timing of bait application: in some areas, farmers report on the rodent outbreak cases and request for control assistance after they observe some cases of crop damage in their fields. This results

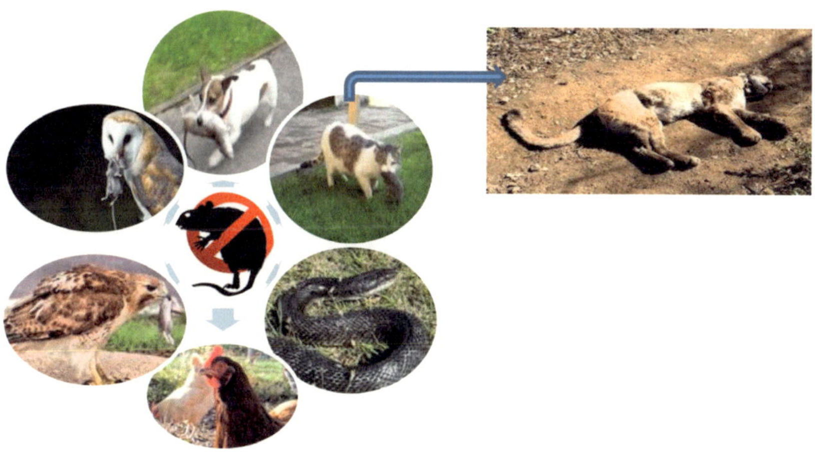

Figure 6.
Effect of acute poison to non-targeted animals. Source: Mdangi [39].

into delayed process in control as it takes time for the information to reach the responsible public rodent control centers. This is critical for alleviating damage [51, 68]. Another factor that limits the use of rodenticides is poverty; many small-scale farmers are poor, and therefore, in many cases, they cannot afford to buy sufficient rodenticides for their farms [23].

In addition, the use of rodenticides and other control methods provides only a short-term solution, and they are not effective in cases of high population as have been reported in irrigated rice systems where rodent breed throughout the year [68, 69]. In order, therefore, to minimize those problems, alternative measures must be sought and one of them being the use of trap barrier system (TBS). It has been reported that the application of TBS could increase yields by 10–25% [8] and is cost-effective in most seasons.

2.2 Philosophy on TBS for rodent management

Trap barrier system is a new environmentally friendly, physical rodent control method. It has been proved very successful in controlling rats in irrigated rice fields in Southeastern Asia.

2.2.1 Construction of TBS

An area of 10 m by 10 m or 20 m by 20 m, which is equal to size of one trap barrier, is constructed and measured by using tape measure, staked and marked with a piece of trees dug 50 cm into the ground and stands for 1.5 m above the ground. String and wire is used to maintain an erect barrier. Thereafter, polythene sheet with size of 45 m length and 1 m width is rolled around the staked pegs/piece of trees followed by covering the sheet with mud below the ground (about 5–10 cm), so that no rodent can penetrate the sheet. Therefore, a significant aspect of the trap barrier system (TBS) is that the crop protection occurs in ecologically acceptable manner, as the entire crop is wrapped in polyethylene sheets and held together with wooden bamboos, at sufficient height of about (90–95 cm) from above ground.

Live-multiple-capture cage traps (240 × 150 × 150 mm) are placed every 2.5 m (n = 8 per TBS) from each angle. The two multiple capture traps are installed along each side inside the sheet held tightly against the fence, facing the hole made on the polythene sheet, making a total of eight holes and eight traps per trap barrier. Trap barrier is repaired for any destruction if occurred (**Figure 7**).

2.2.2 Crop transplanting and animal trapping

The trap (lure crop) is transplanted inside the barrier immediately after trap barrier has been constructed in each season. The seedlings in the surrounding TBS are transplanted 3 weeks later (**Figure 8**). Moreover, every important agronomic practice is done. Trapping in the TBS starts after construction the barrier whereby two multi-capture traps. The multi-capture traps are cleared of rats and re-trap every morning for entire crop growth period.

2.2.3 Potential of TBS in rodent management in Tanzania

2.2.3.1 Rodent pest species captured

Two small mammal species were captured, which included *Mastomys natalensis,* which is a rodent pest species, and *Crocidura* spp., which is an insectivorous species. *Mastomys natalensis* contributed more than 97% of the total small mammals

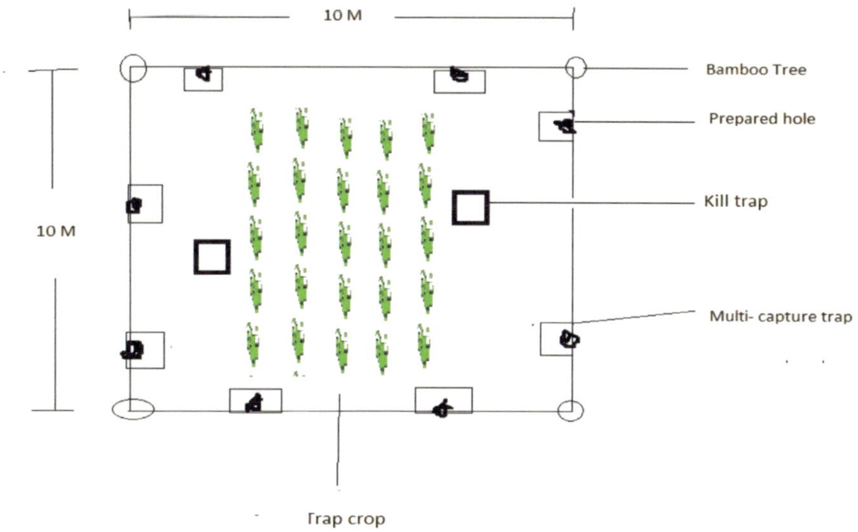

Figure 7.
A pictorial presentation of TBS structure. Source: Courtesy by Loth S. Mulungu.

Figure 8.
Constructed TBS with rice inside planted 3 weeks before planting in surrounding. Courtesy by Grant Singleton.

captured in the study area for both dry and wet seasons. This observation of high abundance of *M. natalensis* is consistent with those reported by Vibe-petersen et al. [70] and Sluydts et al. [71] in maize farms, Makundi et al. [72] and Massawe et al. [73] in fallow fields and by Mulungu et al. [68] in irrigated rice fields. The presence of *M. natalensis* in such high population abundances in this area is probably associated with availability of food, habitat and/or reproduction potential of the species.

According to Makundi et al. [74], the species is a pioneer in colonizing disturbed habitats (e.g. by agriculture). Likewise, Odhiambo et al. [75] and Mulungu et al. [76] reported that the species feeds in almost all types of food in the environment but predominantly prefers seeds/grains. Leirs et al. [63] incriminated *M. natalensis* to be an opportunistic rodent species and named it characteristically to conform with r-selected strategy when conditions are favorable. Rodent populations usually fluctuate from time to time [77].

2.2.3.2 Rodent population reduction

Studies have been conducted in Tanzania to assess the effectiveness of TBS on reduction of rodent pest species population [78]. Results showed no significant effect ($F_{1, 18} = 1.30$; $p = 0.32$) of the TBS on population abundance between dry and wet seasons if TBS has been used regardless the high population abundance of rodent observed on dry season than wet season. During dry season, higher catch ($F_{9, 20} = 9.604$, $p \leq 0.0001$) was observed when the population is higher in October (16.0 animals) (**Table 1**). It has been reported that the fluctuations can be accelerated by factors like food availability and/or other environmental factors such as water flooding or vegetation cover [79, 80]. High population was observed at transplanting and booting stages in dry and wet seasons, respectively, although it is not significant with other crop growth stages. This is contrary with previous observations by Mulungu et al. [68] who reported that high population was recorded during the dry season at transplanting and vegetative crop growth stages.

The discrepancy of these two observations in the same area may be due to a change of planting calendar. Mulungu et al. [68] reported that farmers start land preparation and transplanting in July and January for dry and wet seasons, respectively, whereas in the current study, planting and land preparation starts in July and January for dry and wet seasons, respectively. Generally, in this study, the rodent population decreases with an increase in crop growing stages.

2.2.3.3 Seasonal rodent population reduction

For the wet season, there was a significant interaction effect on monthly and *M. natalensis* population abundance ($F_{9, 20} = 9$, $p \leq 0.0001$) with fields type practice applied. Highest number of rodent catch was observed in May (1.0 animal) than other months (**Table 2**). Lowest trap catches were observed in the control in the month of July (0 animals captured). The increase of crop damage corresponds with the increase of rodent population abundance.

The dry season (October) had high population abundance and high crop damage compared to wet season. This observation concurs with Meheretu et al. [81] in wheat crop who reported that when wheat was at maturity stage, rodent abundance was low. One could expect an increase of population as the crop grows due to availability of shelter and cover. Both the wet and dry seasons are favorable for rat

Month*Management	Mean population	% Damage
Oct*Tbs	16.0 ± 4.1a	27.2331 ± 1.42a
Oct*Control	10.31 ± 1.7a	29.7671 ± 1.96a
Nov*Tbs	2.7 ± 1.2b	20.2000 ± 0.78ab
Nov*Control	2.31 ± 1.2b	21.2672 ± 2.11ab
Dec*Tbs	2.01 ± 1.4b	10.3330 ± 0.79bc
Sept*Control	1.71 ± 1.0b	9.53 ± 7.78bc
Sept*Tbs	1.00 ± 0b	1.0000 ± 0c
Dec*Control	0.00 ± 0b	10.4334 ± 4.88bc
Jan*Control	0.00 ± 0b	3.7003 ± 3.02c
Jan*Tbs	0.00 ± 0b	3.0672 ± 2.5c
DF	9	9
F	5.32	11.680
p	0.0001	<0.0001

Source: Mchukya [78].

Table 1.
Effect of interaction between months and population abundance on dry season.

Month*Management	Mean population	% Damage
May*Tbs	1.00 ± 0.0a	1.000 ± 0.0a
April*Control	0.67 ± 0.54ab	2.033 ± 1.66ab
June*Tbs	0.67 ± 0.54ab	0.700 ± 0.57ab
May*Control	0.67 ± 0.54ab	4.233 ± 1.94ab
April*Tbs	0.50 ± 0.35ab	3.400 ± 0.28ab
June*Control	0.33 ± 0.27b	0.733 ± 0.6b
March*Tbs	0.33 ± 0.27b	0.000 ± 0.0b
July*Control	0.00 ± 0.0b	0.000 ± 0.0b
July*Tbs	0.00 ± 0.0b	0.000 ± 0.0b
March*Control	0 ± 0b	0.967 ± 0.7b
DF	9	9
F	49.977	1.677
p	<0.0001	0.161

Source: Mchukya [78].

Table 2.
Effect of interaction between months and population abundance on wet season.

reproduction and crop damage. The presence of food, water and shelter in the area are factors that permit the survival of rat populations. In rice fields, the quantity and quality of the available harborage usually vary considerably from place to place and season to season. Sumangil [82] reported short-range seasonal movements among *R. argentiven*.

Quick [83] reported that an increase in rice damage towards maturity was associated with an increase in crop cover (i.e. rice tillers) and food (i.e. rice grain). The same was observed by Mulungu et al. [68] who reported that rodent population abundance increases with an increase of rice growth stages. Frequent rains and irrigation, which flooded rat burrows, may have effectively kept rodent activities low or forced some rodents to migrate to domestic environment as *M. natalensis* is semi-domestic species. As observed in wheat fields [84], rat activity increased in fields as the crops matured and the plots became dry.

The occurrence of rodent outbreaks in Tanzania is influenced by the rainfall pattern [85]. Rodents breed during the long rains and usually starts one month after the usual peak rainfall, lasting until dry season [85]. Neonates grow slowly and normally do not mature before the next rainy period. Unless abundant rains appear before March and April the following year, they will be at least 6 months old before they begin to breed [85].

Fulk [86] reported similar influxes of rodents into rice fields in Pakistan. As the rice ripened and water was drained from the plots, rodent numbers increased rapidly. Despite high numbers of rodent individuals recorded at vegetative and booting, rodent damage was lowest at maturity growth stages in both seasons. Average grain yield on the wet season and dry season was not different ($p > 0.05$). Wet season had relatively higher grain yield than the dry season. The lower yield observed during the dry season is probably attributed to rodent damage, irregular irrigation, and/or prolonged periods of water stress caused by insufficient water supply [87].

According to Raes et al. [88], rice cultivated in the dry season experiences much of the moisture stress [89]. Other similar findings include that of Craufurd et al. [90], who reported that water stress has negative impacts on yield and effects vary with phenological stages, which are generally more severe from the flowering stage onwards. Yue et al. [91] reported that yield loss under drought stress could be associated with an increase of spikelet sterility and a reduction in panicle filling rate as well as grain weight. Damage at dry season resulted into lower yield losses compared to wet season.

At early growth stage such as transplanting, yield loss was observed to be higher compared to later growth stages in dry season and vegetative and booting stage at wet season. As damage ascended from zero to 50% stem tiller cut, yield losses followed the same trend. The results of this study also indicate that rice crop damage through the cutting of tillers may have negligible impact on yield, particularly if the damage occurs early in the growing season at the transplanting stage of the crop.

It has been reported that percentage yield loss at these growth stages is roughly approximate to the percentage of damage [92, 93], which is attributed by the fact that at late stages the crop cannot produce more tillers to compensate for damage since very little time is available for such compensatory growth. Compensation in rice crop yield can be further observed through the significant interaction between growth stage and damage level.

The significant interactive effects between growth stage and damage level suggest that rice plant compensation has occurred. Similar findings were reported by Fulk and Akhtar [94] who showed that rice grain yield may not be affected by loss of tillers at their early growth stages as the numbers of productive tillers are determined at the late tillering stage. Buckle et al. [95] reported that compensation capacity of rice damaged by rodents is higher at each growth stage than at maturity of the crop. Aplin et al. [77] explained the term compensation of rice in terms of tiller regrowth and panicle filling.

Cuong et al. [96, 97] observed that the yield loss might be high and probably result in total yield loss when damage occurs at the reproductive phase as there would not be sufficient time for compensation to occur. The difference in grain yield in crop plants could be attributed to the effect of weather, pest pressure (damage), and field management. Average number of panicles per plant in the wet season was observed to be higher than that of the dry season. This perhaps may be due to availability of moisture/flood condition in wet season, which limit rodent movement within the field while others migrate to domestic environment. These results agree well with those of Kim et al. [98] who reported that drought exposure during the earlier stages of reproductive growth affects panicle formation negatively. Also, rodent damage recorded in the dry season was higher than that of the wet season especially plots with no TBS.

2.2.3.4 Radius covered by TBS in controlling rodent pests

Again, Mchukya [78] observed that distance measured (i.e., 0, 10, 20, 30, 50 m) from the TBS differs significantly ($F_{3, 38} = 4.61$; $p = 0.0076$) and indicating that up to 20 m, one structure of TBS manages rodent during dry season. However, at wet season, no effects ($F_{3, 38} = 0.94$; $p = 0.4293$) on rodent abundance between distance were tested. Across the season, there was significant difference between distance ($F_{3, 78} = 4.28$; $p = 0.0075$) where TBS reduces population up to 20 m.

During the dry season, low population and damage were maintained at late stages within a distance of ≤20 m and increased as the distance increased (≥ 20 m) away from the lure crop. On other side of wet season, population and damage were very low at early and late stages but high at vegetative and booting stages. Low population abundance and damage were maintained within a distance of ≤30 m but increased as the distance increased (≥30 m) away from the lure crop.

Across the season, low population and damage were maintained within a distance of 20 m (**Table 3**).

During the dry season, the effect of the TBS was much pronounced within 20 m distance of protection from the trap crop by considering the damage which was very high compared to wet season, although that low damage continue to reduce much more up to 30 m with the aid lure crop within the trap barrier. In this study, the distance covered by TBS was 0.5 acre, which is very small compared to singleton [22], whose TBS was effective within 200 m covering a total area of 15 ha.

The bunds surrounding small plots owned by farmers were acting as home range of rodents, which allow them for easiness of short movement in attacking rice that resulted to a minimum of 20 and 30 m distance of protection from the center of the trap crop in dry and wet seasons, respectively. Trap barrier system has proved very successful in irrigated rice fields in Southeastern Asia to control rats, a cost-beneficial and sustainable solution, and the yield of rice has increased with 10–25%. It is basically a system where rodents are trapped in a rice field that is planted a short period earlier than the surrounding fields and therefore attracting rodents from a much wider area [22].

2.2.3.5 Yield loss

There was no significant difference between treatments within and across the seasons, although the yield over time varied considerably between TBS and field without TBS and seasons. Highest value was observed in plots with TBS than fields without TBS plots in both seasons and across the season. However, there was relatively lower yield in dry season than the wet season due to higher rodent damage and water stress (**Table 4**).

Population abundance corresponds with the increase of crop damage. High population abundance and crop damage were much observed on dry season than wet season, although TBS saved 510 kg of harvested paddy. Assuming that 0.25 kg of rice when cooked can be consumed by one person, this means that a total of 2040

Distance (m)	**Mean damage (%)**		
	Dry season	**Wet season**	**Across season**
0	18.0b	5.0a	0.0667b
10	25.7b	9.9a	0.333ab
20	28.2ab	15.1a	0.5333ab
30	51.3a	25.3a	0.7333a
Source: Mchukya [78].			

Table 3.
Crop damage (%) at different distances within and across the seasons.

Treatment	**Yield (t/ha)**		
	Dry season	**Wet season**	**Across season**
Fields with TBS	3.83	5.69	4.76
Fields without TBS	3.323	4.33	3.83
Source: Mchukya [78].			

Table 4.
The yield of rice (t/ha) obtained from plots with TBS and those without TBS within the seasons.

people per meal in a given area or village could benefit from system. The cost-benefit ratios for the dry and wet seasons, respectively, indicate the strong potential of a TBS with trap crop for managing the rice field rat.

2.2.3.6 Economics on the use of trap barrier system

The benefits from all fields with TBS during wet and dry season were relatively high compared with that of the fields without TBS. Fields with TBS had higher undamaged tillers, which resulted in the increase of revenues that exceeded the cost of the plant protection regime, although it was noticed that the cost of plant protection using TBS was higher than fields without TBS. The yield from TBS and fields without TBS plots were 3.83 and 3.323 t/ha in the dry season (**Table 5**) and 5.69 and 4.33 t/ha in wet season (**Table 6**), respectively. Across the season (**Table 7**), the fields with TBS had higher mean yield (4.76 t/ha) compared to fields without TBS (3.83 t/ha). The benefit was obtained by taking the yield (t/ha) multiply by 900 Tsh/kg of harvested paddy. Therefore, the cost-benefit ratios for using a TBS were

Fields types	Yield (kg/ha)	Increased yield over control	Value of yield (Tsh)	Materials, labor, bait, rodenticides	Net benefit (NB)	Cost-benefit ratio (CBR)
With TBS	3830	507	456,300	215,000	241,300	**1:1.1**
Without TBS	3323					

Source: Mchukya [78].

Table 5.
Evaluation of the cost and benefit of control rodent pests with trap barrier system in dry season.

Fields types	Yield (kg/ha)	Increased yield over control	Value of yield (Tsh)	Materials, labor, bait, rodenticides	NB	CBR
With TBS	5690	1360	1,224,000	160,000	1,064,000	**1:6.7**
Without TBS	4330					

Source: Mchukya [78].

Table 6.
Evaluation of the cost and benefit of control rodent pests with trap barrier system in wet season.

Fields types	Yield (kg/ha)	Increased yield over control	Value of yield (Tsh)	Materials, labor, bait, rodenticides	NB	CBR
With TBS	4760	934	840,600	187,500	653,100	**1:3.5**
Without TBS	3826					

Source: Mchukya [78].

Table 7.
Evaluation of the cost and benefit of control rodent pests with trap barrier system across the seasons.

1:1.1 for the dry season, 1:6.7 for the wet season and 1:3.5 across the season. This is in contrast to the use of a TBS alone which, in Malaysia and the Philippines, requires crop losses of >30% before there is a positive benefit-cost ratio [99]. There has been only one report in Southeast Asia of high benefit-cost ratios for a TBS alone: ratios of 19:l and 28:l in Malaysia in a region where 56% of rice farms had suffered yield losses [100]. Murakami [101] also reported a TBS to be effective against *R. qyntil-*water in paddies that had severe rat damage during the previous year.

The main factor providing the high benefit-cost ratio is the halo of protection provided to crops outside the TBS. Therefore, the selection of the project with the benefit-cost ratio or Profitability index (PI) method can also be done on the basis of ranking. The highest rank will be given to the project with the highest PI, followed by the others in the same order. According to Misuraca [102], the cost-benefit ratio exceeding one might be termed as the project worth undertaking as it become comparable to increasing returns to scale contributed by the project if a firm adopt it. The higher the cost-benefit ratio results in the higher net return [20]. The effect of a TBS plus trap crop on mean yield increased up to 20 m from the TBS and the associated cost-benefit ratios in this study.

2.3 Conclusion

Trap barrier system indicated its strong potential in lowering population abundance at a distance within 20 m away from the trap, which corresponding with low damage resulted to high yield. It is therefore recommended the small-scale farmers to use trap barrier system against pre-harvest rat losses to rice probably during dry season due to high rat densities. This will help them to save more, maximize their profit, and improve their living standard. Also, further study is needed to test this new technology in other irrigation schemes and other African countries and to determine the home range of rodent in order to provide valuable comparative data basis.

Author details

Loth S. Mulungu[1]*, Bernard M. Mchukya[2] and Laurent L. Mnyone[1]

1 Pest Management Centre, Sokoine University of Agriculture, Morogoro, Tanzania

2 Rodent Control Centre, Ministry of Agriculture, Food Security and Cooperatives, Morogoro, Tanzania

*Address all correspondence to: lothmulungu@yahoo.co.uk

References

[1] International Rice Research Institute (IRRI). Rice Knowledge Bank. 2009. Available from: www.irri.org [Accessed: 01-07-2017]

[2] Banwo O. Management of major insect pest of rice. In: Makundi RH, editor. Morogoro, Tanzania: Pest Management Centre, Sokoine University of Agriculture & Dar es Salaam: Tanzania Publishing House LimitedManagement of Selected Crop Pest of Tanzania. 2006. pp. 66-79

[3] Wayne SC, Henry DR. Rice: Origin, History, Technology, and Production. Hoboken, NJ: John Wiley & Sons. Wildlife Research. 2003;**37**:355359

[4] Nguyen VN, Labrada R. FAO Rice Information. Vol. 3. Rome: Food and Agriculture Organisation of the United Nations; 2002

[5] Buckle AP. Rodent control methods: Chemical. In: Buckle AP, Smith RH, editors. Rodent Pests and Their Control. Wallingford: CAB International; 1994. pp. 127-160

[6] Fall MW. Rodents in Tropical Rice. Technical Bulletin No. 36. University of the Philippines at Los Banos; 1977. p. 39

[7] Wood BJ. Rodents in agriculture and forestry. In: Buckle AP, Smith RH, editors. Rodent Pests and Their Control. Wallingford, Oxon, UK: CAB International; 1994. p. 405

[8] Singleton GR, Steve R, Peter B, Brown R, Hardy B. Rodent Outbreaks: Ecology and Impacts. Los Baños (Philippines): International Rice Research Institute; 2010. p. 289

[9] Meerburg BG, Singleton GR, Leirs H. The year of rat ends—Time to fight hunger! Pest Management Science. 2009;**65**:351-352

[10] Massawe WA, Mulungu LS, Makundi, Mulungu LS. Vertebrate pest and their management: A general overview. In: Makundi RH, editor. Management of Selected Crop Pests in Tanzania. Dar Es Salaam, Tanzania: Tanzania Publishing House Limited; 2006. pp. 236-247

[11] Mulungu LS, Makundi RH, Leirs H, Massawe A, Vibe-Petersen S, Stenseth NC. The rodent density—Damage function in maize fields at early growth stage. In: Singleton GR, Hinds LA, Krebs CJ, Spratt DM, editors. Rats, Mice and People: Rodent Biology and Management Singleton. Canberra: ACIAR; 2003. pp. 301-313

[12] Singleton GR, Belmain SR, Brown PR. Rodent outbreaks: An age-old issue with a modern appraisal. In: Singleton GR, Belmain SR, Brown PR, Hardy B, editors. Rodent Outbreaks: Ecology and Impacts. Los Banos (Philippines): International Rice Research Institute; 2010. p. 289

[13] Reissig WH, Heinrich EA, Lit singer JA, Moody K, Fiedler L, Mew TW, et al. Illustrated Guide to Integrated Pest Management in Rice Tropical Asia. International Rice Research Institute (IRRI). Philippines; 1985. p. 411

[14] Brown PR, Singleton GR. Impacts of House Mice on Crops in Australia—Cost and Damage. Human Conflicts with Wildlife: Economic Considerations Paper 6. Lincoln: University of Nebraska; 2000. pp. 48-58

[15] Meerburg BG, Singleton GR, Leirs H. The year of the rat ends time to fight hunger! 2008

[16] Singleton GR, Belmain SR, Brown PR, Aplin K, Htwe NM. Impacts of rodent outbreaks on food security in Asia. 2010

[17] Sixbert V. Population dynamics and breeding patterns of rodent pest species and crop loss assessment in irrigated rice in Mvomero district [MSc dissertation]. Morogoro, Tanzania: Sokoine University of Agriculture. 2013. p. 72 [submitted for partial fulfillment of the requirement for the Degree of Masters of science in Crop Science]

[18] Brown PR, Huth NI, Banks PB, Singleton GR. Relationship between abundance of rodents and damage to agricultural crops. Journal of Agriculture, Ecosystem and Environment. 2006;**120**:405-415

[19] Gergon BE, Catudan BM, Desamero NV. Ecology based rat management system in Banaue and Hungduan rice terraces. In: Singleton GR, Joshi RC, Sebastian LS, editors. Philippines rats: Ecology and management. Luzon: Philippine Rice Research Institute; 2008. pp. 85-100

[20] Junaid NAAZM, Ali A. Economic analysis based on benefit cost ratio approach for rice varieties of Malakand division. Economic Analysis. 2005;**6** (19)

[21] Mulungu LS, Makundi RH, Massawe AW. Characteristics of rodent damage to major crops in Tanzania and crop loss assessment techniques. In: Makundi RH, editor. Management of Selected Crop Pests in Tanzania. Dar Es Salaam, Tanzania: Tanzania Publishing House Limited; 2006. p. 476 (ISBN: 9976 1 0210 0)

[22] Singleton GR, Leirs H, Hinds LA, Zhang Z. Ecologically-based management of rodent pest-re-evaluating our approach to an old problem. In: Singleton GR, Leirs H, Hinds LA, Zhang Z, editors. Ecologically-based management of rodent pests. ACIAR Monograph No. 59. Canberra: Australia Centre for International Agriculture Research; 1999. pp. 17-30

[23] Mulungu LS, Sixbert V, Ngowo V, Mdangi M, Katakweba AS, Tesha P, et al. Spatiotemporal patterns in the distribution of the multimammate mouse, Mastomys natalensis in rice crop and fallow land habitats in Tanzania. Mammalia. 2015;**79**:177-184

[24] Mulungu LS, Lopa H, Mashaka Mdangi E. Comparative study of population dynamics and breeding patterns of *Mastomys natalensis* in system rice intensification (SRI) and conventional rice production in irrigated rice ecosystems in Tanzania. Journal of Rice Research. 2016;**4**:161. DOI: 10.4172/2375-4338.1000161

[25] Jacob J. Ecologically-based management of pest rodents in rice-based agro-ecosystems in southeast Asia. In: Proceedings of the 20th Vertebrate Pest Conference. Davis, California, USA: University of Davis; 2002

[26] My Phung NT, Brown PR, Leung LK-P, Tuan LM. The effect of simulated rat damage on irrigated rice yields and compensation. Crop Protection. 2010; **29**:14661471

[27] Taylor D. An outbreak of rats in agricultural areas of Kenya in 1962. Journal of East African Agriculture. 1968;**34**:66-77

[28] Brown PR. Relationship between abundance of rodents and damage to agricultural crops. Agriculture, Ecosystems & Environment. 2007; **120**(2):405-415

[29] Lurz R. An epidemic of plague on Kilimanjaro in 1912. In: Kilonzo BS, Mhina J, Sabuni C, Mgole G, editors. The Role of Rodents and Small Carnivores in Plague Endemicity in

Tanzania. Belgian Journal of Zoology. 1913:**135**:119-125

[30] Mwanjabe PS, Sirima FB, Lusingu JR. Crop losses due to outbreak of *Mastomys natalensis* (Smith, 1834). Muridae, Rodentia, in Lindi region Tanzania. International Biodeterioration and Biodegradation. 2002;**49**(2):133-137

[31] Mulungu LS, Ngowo VD, Makundi RH, Massawe AW, Leirs H. Winning the fight against rodent pests: Recent developments in Tanzania. Journal of Biological Sciences. 2010;**10**:333-340

[32] Ngowo V, Lodal J, Mulungu LS, Leirs H, Makundi RH, Massawe AW. Evaluation of Thiram and Cinnamamide as potential repellents against maize seed depredation in the field by the multimammate rat, *Mastomys natalensis*, in Tanzania. In: Singleton GR, Hinds LA, Krebs CJ, Spratt DM, editors. Rats, Mice and People: Rodent Biology and Management. ACIAR Monograph No. 96. 2003. p. 564

[33] Mulungu LS, Sabuni CA, Ngowo V, Mdangi ME, Magadula AJ, Kapingu M, et al. Domestic cat (*Felis silvestris catus*) urine odour as a potential rodent management strategy. Huria. 2016;**22**: 144-152

[34] Massawe AW, Leirs H, Rwamugira WP, Makundi RH. Effect of land preparation methods on spatial distribution of rodents in crop fields. In: Singleton GR, Hinds LA, Krebs CJ, Spratt DM, editors. Rats, Mice and People: Rodent Biology and Management. Vol. 564. Canberra: Australian Centre for International Agricultural Research; 2003. pp. 229-232

[35] Masol JR, Epple G, Nolte DL. Semiochemicals and Improvement in Rodent Control. Philadelphia, USA: U.S. Department of Agriculture, Animal and Plant Health Inspection Service; 1994. pp. 327-345

[36] Voznessenskaya VV, Wysocki CJ, Zinkevich EP. Regulation of the rat oestrous cycle by predator odors: Role of the vomeronasal organ. In: Doty RL, Muller-Schwarze D, editors. Chemical Signals in Vertebrates. New York: Plenum Press; 1992. pp. 281-284

[37] Voznessenskaya VV, Naidenko SV, Feoktistova NY, Krivomazov GJ, Miller LA, Clark L. Predator Odors as Reproductive Inhibitors for Norway Rats. USDA National Wildlife Research Centre – Staff Publications; 2003, paper 251. Available from: http://digitalc ommons.unl.edu/icwdm_usdanwrc [Accessed: 11-07-2016]

[38] Mulungu LS, Sabuni CA, Ngowo V, Mdangi ME, Magadula AJ, Kapingu M, et al. Effectiveness of the domestic cat (*Felis silvestris catus*) urine extracts odour against commensal rodents. Huria. 2017

[39] Mdangi ME. Evaluation of selected pesticidal plants for maize (*Zea mays* L.) seed protection against *Mastomys natalensis* at sowing and early seedling stage [PhD thesis]. Morogoro, Tanzania: Sokoine University of Agriculture. 2017. p. 179 [submitted for partial fulfillment of the requirement for the Degree of Doctor of Philosophy]

[40] Fiedler LA, Fall MW. Rodent control in practice: Tropical field crops. In: Buckle AP, Smith RH, editors. Rodent Pests and Their Control. Wallingford, Oxon, UK: CAB International; 1994. pp. 313-338

[41] Day T, MacGibbon R. Multiple-species exclusion fencing and technology for mainland sites. New Zealand: Proof fencing ltd, Cambridge university; 2007. pp. 418-433

[42] Sridhara S. Vertebrate Pest in Agriculture, The Indian Scenario. India: Scientific Publishers; 2006. p. 605

[43] Makundi RH, Bekele A, Leirs H, Massawe AW, Rwamugira W, Mulungu LS. Farmer's perceptions of rodents as crop pests: Knowledge, attitudes and practices in rodent pest management in Tanzania and Ethiopia. Belgium Journal of Zoology. 2005;**135**:153-157

[44] Sarker NJ, Rokunuzzam MD, Nessa RR. Abundance of rats and mice in the selected areas of Dhaka city: A cross sectional study. Journal of Entomology and Zoology Studies. 2013;**1**(5):116-119

[45] Witmer GW, Moulton RS, Swartz JL. Rodent burrow systems in North America: Problem posed and potential solution. In: Timm RM, editor. Proceeding of the 25th vertebrate Pest conference. Davis: University of California; 2012. pp. 208-212

[46] Thankur NSA, Firake DM, Kumar D. Indigenous traps for the management of rodent outbreak in north eastern hill region of India. Indian Journal of Traditional Knowledge. 2013;**12**:730-735

[47] Bale JS, van Lenteren JC, Bigler F. Biological control and sustainable food production. Journal of Philosophical Transactions of the Royal Society. 2008; **363**:761-776

[48] Singleton GR, Redhead TD. Structure and biology of house mouse populations that plague irregularly: An evolutionary perspective. Biological Journal of the Linnean Society. 1990;**41** (1–3):285-300

[49] Howard WE. Biological control of vertebrate pests. In: Proceedings of the 3rd Vertebrate Pest Conference. 1967

[50] Nelson J. Effectiveness of rodent control and monitoring techniques for a montane rainforest. Wildlife Society Bulletin. 2002:82-92

[51] Tobin ME, Sugihara RT, Ta AJ. Rodent damage to Hawaiian sugarcane. In: Proceedings of the Vertebrate Pest Conference. Vol. 14. 1990. pp. 120-123

[52] Hafidzi MN, Mohd NI. The use of the barn owl, Tylo alba, to suppress rat damage in rice fields in Malaysia. ACIAR Monograph Series. 2003;**96**: 274-276

[53] Taylor KA. Vertebrate pest problems in Africa. In: Dubock AC, editor. Proceedings of a Conference on: The Organization and Practice of Vertebrate Pest Control. Surrey, England: Imperial Chemical Industries PLC; 1984. p. 662

[54] Cox GW. Alien Species in North America and Hawaii, Impacts on Natural Ecosystems. Washington, D. C: Island Press; 1999. p. 387

[55] Baldwin PH, Schwartz CW, Schwartz ER. Life history and economic status of the mongoose in Hawaii. Journal of Mammalogy. 1952;**33**:335-356

[56] Kami HT. Foods of the mongoose in the Hamakua district, Hawaii. Zoonoses Research. 1964;**3**(3):165-170

[57] Bateman JA. Animal Traps and Trapping. Harrisburg, PA, USA: Stackpole Books; 1971. p. 286

[58] Bateman JA. Trapping, A Practical Guide. Harrisburg, PA, USA: Stackpole Books; 1979. p. 168. Available from: Knowledgebank.irri.org [Accessed: 28-11-2017]

[59] Fall MW, Jackson WB. A new era of vertebrate pest control? An introduction. International Biodeterioration & Biodegradation. 1998;**42**(2):85-91

[60] Gosling LM, Baker SJ. The eradication of muskrats and coypus from Britain. Biological Journal of the Linnean Society. 1989;**36**:39-51

[61] Makundi RH, Oguge NO, Mwanjabe PS. Rodent pest management in East Africa—An ecological approach. In: Singleton G, Leirs H, Zhang Z, Hinds L, editors. Ecological-Based Management of Rodent Pest. Canberra, Australia: ACIAR; 1999. pp. 460-476

[62] Ngowo V, Lodal J, Leirs H, Massawe AW, Makundi RH. Evaluation of thiram and cinnamamide for maize seeds protection against multimammate rat, *Mastomys natalensis*, in maize fields in Tanzania. Belgian Journal of Zoology. 2005;**135**(Suppl):179-181

[63] Leirs H, Stenseth N, Nichols JD, Hines JE, Verhagen R, Verheyen W. Stochastic seasonality and non-linear dependent factors regulate population size in an African rodent. Nature. 1997; **389**:176-180

[64] Buckle AP. Rodenticide—Their role in rodent pest management in tropical agriculture. In: Singleton GR, Hinds LA, Leirs H, Zhang Z, editors. Ecologically-Based Management of Rodent Pests. Canberra, Australia: Australian Centre for International Agricultural Research; 1999. pp. 163-177

[65] Gratz NG. A critical review of currently used single-dose rodenticides. Bulletin of the World Health Organization. 1973;**48**(4):469

[66] Bai KM. Effects of the addition of L-histidine on the toxicity of zinc phosphide in laboratory rats and mice. ACIAR Monograph Series. 2003;**96**: 477-480

[67] Pratt H. Rodenticides: What to use where, when and how. Pest Control. 1983;**51**(10):19-26

[68] Mulungu LS, Ngowo V, Mdangi M, Katakweba AS, Tesha P, Mrosso FP, et al. Population dynamics and breeding patterns of multimammate mouse, *Mastomys natalensis* (Smith 1834) in irrigated rice field in eastern Tanzania. Pest Management Science. 2013;**69**: 371-377

[69] Mulungu LS. Control of rodent pests in maize cultivation: The case of Africa. In: Watson D, editor. Achieving Sustainable Maize Cultivation. Vol. 2. Francis Dodds, Burleigh Dodds Science Publishing; 2017. DOI: 10.199103/AS.2016.0002.18

[70] Vibe-Petersen S, Leirs H, Bruyn LD. Effects of predation and dispersal on *Mastomys natalensis* population dynamics in Tanzanian maize fields. Journal of Animal Ecology. 2006;**75**: 213-220

[71] Sluydts V, Davis S, Mercelis S, Leirs H. Comparison of multimammate mouse (*Mastomys natalensis*) demography in monoculture and mosaic agricultural habit: Implication for pest management. Journal of Crop Protection. 2009;**28**:647-654

[72] Makundi RH, Massawe AW, Mulungu LS, Katakweba A. Species diversity and population dynamics of rodents in a farm-fallow field mosaic system in central Tanzania. African Journal of Ecology. 2009;**48**:313-320

[73] Massawe WA, Mulungu LS, Makundi RH, Dlamini N, Eiseb SJ, Kirsten F, et al. Spatial and temporal population dynamics of rodents in three geographically different regions in Africa: Implication for ecologically-based rodent management. African Journal of Zoology. 2011;**46**:393-405

[74] Makundi RH, Massawe AW, Mulungu LS. Reproduction and population dynamics of *Mastomys natalensis* Smith, 1834 in an agricultural landscape in the Western Usambara Mountains, Tanzania. Integrative Zoology. 2007;**2**(4):233-238

[75] Odhiambo RO, Makundi RH, Leirs H, Verhagen R. Dietary selection in *Mastomys natalensis* (Rodentia: Muridae) in the maize agro-ecosystems of central

and southwestern Tanzania. Mammalia. 2005;**72**:169-177

[76] Mulungu LS, Mahlaba T'a A, Massawe AW, Kennis J, Crauwels D, Eiseb S, et al. Dietary differences of the multimammate mouse, *Mastomys natalensis* (smith 1834), across different habitats and seasons in Tanzania and Swaziland. Wildlife Research. 2011; **38**(7):640-646

[77] Aplin KP, Brown PR, Jacob J, Krebs CJ, Singleton GR. Field Methods for Rodent Studies in Asia and the IndoPacific. ACIAR Monograp. 100. Canberra (Australia): Australian Centre for International Agricultural Research; 2003

[78] Mchukya BM. Evaluation of trap barrier system as a rodent management tool in irrigated rice fields [MSc Dissertation]. Morogoro, Tanzania: Sokoine University of Agriculture. 2018. p. 108 [submitted for partial fulfillment of the requirement for the Degree of Crop Science]

[79] Douangboupha B, Brown PR, Khamphoukeo K, Aplin KP, Singleton GR. Population dynamics of rodent pest species in upland farming systems of LAO PDR. Kasetsart Journal (Natural Science). 2009;**43**:125-131

[80] Mulungu LS, Ngowo V, Mdangi M, Katakweba AS, Tesha P, Mrosso FP, et al. Survival and recruitment of the multimammate mouse, *Mastomys natalensis* (Smith 1834), in a rice agro-ecosystem. Mammalia. 2015. DOI: 10.1515/mammalia-2014-0150

[81] Meheretu Y, Sluydts V, Welegerima K, Bauer H, Teferi K, Yirga G, et al. Rodent abundance, stone bund density and its effects on crop damage in the Tigray highlands, Ethiopia. Crop Protection. 2014;**55**:61-67

[82] Sumangil JP. Burrows and breeding of ricefield rats in Cotabato, Philippines. Proc. Pest Control Counc. Philippines. 1972;**3**:61

[83] Quick GR. Report and proceedings of an expert meeting on rice rodent control; 10-14 September 1990; Los Banos, Philippines. 1990

[84] Poche RM, Mian MY, Haque ME, Miah D, Mannan A (I), Mannan A (II), et al. Rodent Damage to Wheat in Bangladesh. Special Report No. 1. Bangladesh Agricultural Research Institute, Vertebrate Pest Division; 1979. p. 19

[85] Leirs H. Population ecology of *Mastomys natalensis* multimammate rats: Possible implications for rodent control in Africa. Agricultural Editions. 1995;**35**:273

[86] Fulk G. Food hoarding of *Bandicota bengalensis* in a rice field. Mammalia. 1977;**41**:59-61

[87] Nguyen VN, Ferrero A. The sustainable development of rice-based production systems in Europe. International Relations Council (IRC) Newsletter. 2004;**54**:115-124

[88] Raes D, Kafiriti EM, Wellens J, Deckers J, Maertens A, Mugigo S, et al. Can soil bunds increase the production of rain-fed lowland rice in south eastern Tanzania? Agricultural Water Management. 2007;**89**:229-235

[89] Sumarno S, Sutisna E. Identification of rice (*Oryza sativa* l.) varieties suitable for dry season and wet season planting. Indonesian Journal of Agricultural Science. 2010;**11**:24-31

[90] Craufurd PQ, Vadez V, Jagadish SVK, Prasad PVV, Allaha Z. Crop science experiments designed to inform crop modeling. Agricultural and Forest Meteorology. 2013;**170**:8-18

[91] Yue B, Xue W, Xiong L, Yu X, Luo L, Cui KK, et al. Genetic basis of drought resistance at reproductive stage in rice: Separation of drought tolerance

from drought avoidance. Genetics. 2006;**172**(2):121312

[92] Singleton GR. Impacts of rodents on rice production in Asia. Discussion Paper Series No. 45. Los Banos, Philippines: International Rice Research Institute. p. 30

[93] Poche RM, Haque ME, Mian MY, Sultana P, Karim MA. Rice yield reduction by simulated rat damage in Bangladesh. Journal of Tropical Pest Management. 1981;**27**:242-246

[94] Fulk GW, Akhtar MT. An investigation of rodent damage and yield reduction in rice. Tropical Pest Management. 1981;**27**(1):116120

[95] Buckle AP, Yong YC, Rowe FP. Yield response of the rice variety improved Mahsuri to simulated rat damage. Malaysian Agricultural Journal. 1979;**52**:135-144

[96] Cuong LQ, Chien HV, Han LV, Duc VH, Singleton GR. Relationship between rodent damage and yield loss in rice in Mekong Delta. In: Singleton GR, Hinds LA, Krebs CJ, Spratt DM, editors. Rats, Mice and People: Rodent Biology and Management. Canberra: ACIAR; 2003. pp. 297-300

[97] Kubo M, Purevdorj M. The future of rice production and consumption. Journal of Distribution Research. 2004; **35**(1):128-142

[98] Kim HE, Park HS, Kim KJ. Methyl jasmonate triggers loss of grain yield under drought stress. Plant Signaling & Behavior. 2009;**4**:348-349

[99] Singleton GR, Petch DA. A Review of the Biology and Management of Rodent Pests in Southeast Asia. ACMR7khnical Reports 30. Canberra: Australian Centre for International Agricultural Research; 1994

[100] Lam M-Y. A Review of Food Research in Vietnam, with Emphasis on Postharvest Losses. No. 113896. Australian Centre for International Agricultural Research; 1993

[101] Murakami O, Tiandra Kirana VL, Privono J, Triastiani H. Tikus Sawah. Fiml Report jrlrlonc~.sicr,l-J~Joiut Progrwnnw OH Food Crop Protc~ctiorl Projcw (A M - 162) I'l~usc II. Indonesia: Directorate of Food Crop Protection; 1992. p. 101

[102] Misuraca P. The effectiveness of a costs and benefits analysis in making Federal Government decisions: A literature review. IGARSS. 2014; **2014**(1):1-5. DOI: 10.1007/s13398-014-0173-7.2

Chapter 4

Stem Borers of Cereal Crops in Africa and Their Management

Abou Togola, Ousmane Boukar, Manuele Tamo and Siva Chamarthi

Abstract

The economic importance of the stem borer in Africa results in their severe damage that affect directly cereal yield factors such as the density of fertile tillers and the number of effective panicles. The objective of this paper is to describe and discuss the management options of the main prevalent stem borer of cereal crops in Africa. Host plant resistance, cultural practices, biological control and reasoning chemical control are among the most encouraging options. Integrated pest management combining several compatible methods was highlighted as the most sustainable control option. This paper will served as support for the current research on cereal crops but also as relevant prospect document for entomologists and breeders from across the world.

Keywords: IPM, environment friendly, yield loss, biocontrol, cultural practices

1. Introduction

Stem borers constitute the most widely distributed and injurious group of insect pests of cereal crops. They are commonly known to be one of the limiting factors of cereal production worldwide. They are present in field throughout the crop growing stage from seedling to maturity. The stem borers found on cereal crops in Africa are mainly lepidopterans and dipterans. Cereal crops such as rice, sorghum, maize, sugarcane and pearl millet suffer from the attack of stem borers. The larval stage constitutes the most damaging developmental stage of the pest. They are concealed inside the stem where they feed on the internal cavity of the plant making them very difficult to control. Stem borers cause severe damage on plant stems particularly the destruction of the central leaves (dead-heart) and the drying of the panicle (white head). Their attack leads to significant yield losses. According to [1], the most serious pests of cereal crops in Africa include stem borers. The severity of damage depends not only on the species and density of the pest but also on the phenology stage of the crops. Yield losses of about 10 and 100% due to stem borer have been recorded in rice fields [2].

To control these pests, various strategies have been practiced. These include cultural practices, host plant resistance, habitat management, biocontrol and the use of synthetic pesticides. Each management method has some advantages and limitation regarding its impact on environment, human health and its economic costs and sustainability. Nowadays, the integrated management, combining two or several of these management methods appears to be the most effective and sustainable option.

This chapter comprises two main sections. The first section gives an overview of the main stem borers and their host range in Africa and the second section describes the various management options used to control stem borers and discusses the advantage and limitation of each method while exploring option of combining multiple methods to sustainably mitigate the effect of the stem borer on stakeholder farmers in Africa. This chapter will support the current research on the sustainable management of stem borers attacking cereal crops and will contribute to increase their productivity in Africa.

2. Overview of the main stem borers of cereal crops in Africa

Several stem borers species have been reported to cause severe damage on various cereals crop in Africa.

The maize stalkborer, *Busseola fusca* (Lepidoptera: Noctuidae) is reported to be of economic importance for maize and sorghum [3] while maintaining its population on some alternative hosts. The spotted stem borer, *Chilo partellus* (Lepidoptera: Pyralidae) is considered to be one of the devastating stem borer of sorghum and maize [4, 5] and also makes severe damage on rice in some African countries (Togola, unpublished data). The African striped rice borer, *Chilo zacconius* Bleszynski (Lepidoptera: Pyralidae) is among the major rice stem borer species occurring in the humid forest and savanna zones according to Akinsola [6]. The host range comprises cultivated rice, wild rice, *Oryza longistaminata*, *Panicum* sp., and *Paspalum scrobiculatum* but it has been found also on maize [7]. Other host plants include *Pennisetum* spp., *Rottboellia cochinchinensis* (Loureiro) W.D. Clayton, *Saccharum officinarum* L and *Sorghum arundinaceum* (Desv.) Stapf. [6, 8, 9]. The pink Stem Borer *Sesamia calamistis* (Noctuidae) is generally less important than *Busseola fusca* and *Chilo partellus* as a pest of cereal crops in Africa but may be locally abundant. It attacks sorghum, maize, rice and sugarcane as main host. It can be found also on wheat and pearl millet.as secondary crop host and at less extend on wild grass such as *Pennisetum purpureum, Setaria* sp., *Rottboellia exaltata* and *Cyperus distans as alternative host* [10, 11]. The millet stem borer *Coniesta ignefusalis* Hampson (*Lepidoptera*: *Pyralidae*) is an important pest of pearl millet (*Pennisetum glaucum* (L.) especially in West Africa [12]. The damage caused by *C. ignefusalis* is estimated to 15 and 100% of crop losses annually, depending on location and season [13]. The last generation enters diapause at the beginning of the dry season and stay for 6 months the time the next growing season comes. The sugarcane stem borer *Eldana saccharina* (Pyralidae) is a stem borer of cereal crops in Africa with particular economically importance on sugarcane. In the past *E. saccharina* appeared to be of very little important pest in Africa, except on sugarcane. But it has recently increased in importance on other crops such as maize, and sorghum in several African countries [14]. Also it can attack rice. Its hosts among wild grasses are *Panicum maximum*, *Cyperus papyrus* [15], *Sorghum halepense, S. verticilliflorum and Pennisetum purpureum*. The white rice borer *Maliarpha separatella* Ragonot (Lepidoptera: Pyralidae) is an important stem-borer of rice in West Africa [16]. The larva bores into the stem from the lowest internode where it feeds on the internal tissue preventing the nutrient to rise up until the panicle. The damage caused by this stem borer is said to be unique among rice stem borers because it rarely causes deadhearts or whiteheads [7]. The symptoms of *M. separatella* damage are similar to that of the sheath rot caused by a fungus pathogen *Sarocladium oryzae*. The stem of the infested plant becomes weakened, the panicles incompletely exerted from the flag leaf and the grains incompletely filled with brown coloration. *M. separatella* is more severe in low land, irrigated and floated rice. It also attacks the wild rices *Oryza longistaminata* and *Oryza punctata* [17]. It was also reported on some wild grasses such as *Andropogon tectorum* and *Echinochloa holubii* [18]. The rice yellow stem borer *Scirpophaga spp* (Lepidoptera: Pyralidae) are among the minor rice stem borers in Africa. Several species of *Scirpophaga* exist but the most dominant in Africa is *S. melanoclista* Meyrick [7]. The stalk-eyed fly *Diopsis spp*. (Diptera: Diopsidae) is a serious pest of rice in Africa. The two main species commonly found in rice in Africa are *Diopsis thoracica* Westwood and *Diopsis apicalis* Westwood [7]. Diopsid can be found in all rice ecological zones but preferentially in humid and shady lowland [19–21] and also in irrigated rice fields [7]. Damage from Diopsid larvae is similar to the primary damage made by Lepidopteran larvae resulting to the death of the central leaf of rice plant (deadheart). Feeding by the larvae significantly reduces the tiller density, the effective panicles, the grains weight and the total yield [7] and increases the number of immature panicles. The damage level increases according to *Diopsis* density. In endemic area 60% of the tillers can be infested [22]. Finally the African rice gall midge (AfRGM), *Orseolia oryzivora* Harris and Gagné (Diptera: Cecidomyiidae) is an indigenous dipteran borer of rice that was first reported from southern Sudan in 1947 [23, 24]. The pest is now spread in more than 20 African countries where severe yield losses have been reported. The damage converted the shoot meristem into a gall. The infested plant is no longer able to develop into a floral meristem and then the reproductive potential of the plant is severely compromised [23]. Larval feeding causes severe damage to rice during the vegetative stages (seedling to panicle initiation). Heavy yield losses of 45–80% in farmers' rice crops have been recorded in some fields [25, 26].

A clear knowledge of these stem borer species and their host crop are of key importance for a sustainable management action.

3. Management of the main stem borers of cereal crops in Africa

Because of the nature of the habitat of stem borers (internal shelter), their management requires some specific control measures and actions. Various strategies exist for managing stem borers' population and damage in cereals crops. These include cultural practices, host plant resistance, biocontrol and use of synthetic pesticides.

3.1 Preventive cultural practices

Cultural practices are considered as classic pest control methods. This method consist of manipulating the cropping systems in order (1) to avoid the meeting of crop susceptible stage with pest highest density or (2) to improve the crop growing condition or (3) to make the environment unfavorable for pest proliferation. The cultural practices have the advantage to be easy to implement with less cost. They are more convenient for smallholder farmers in developing countries [27]. Preventives cultural practices comprise a wide range of agronomic practices. These tactics need to be undertaken as first line defense measures to prevent high infestation of stem borers in cereals fields. Among the most effectives cultural practices in controlling stem borers there is cereals intercropping or strip cropping with non-host crops such as cowpea, soybeans and groundnut. Also the choice of appropriate date for planting cereals crops allows the crops to escape to critical period where the pest pressure is high [28]. The sol fertilization and field hygiene are cultural practices that reinforce the plant vigor and increase its defense system. [29, 30] demonstrated that zinc fertilization and potassium fertilization significantly decrease stem borers population in rice and increase paddy yield. Other practice such as destruction of crop residues (burning, plowing or disking) appears to be an effective cultural tactic for limiting the number of diapausing larva of stem borer. [31] demonstrated that plowing and disking crop residues destroyed 24% of the stem borers' population on sorghum and 19% of maize stem borers. Similarly, [32] reported that the destruction of sugarcane residue after harvest significantly reduced the infestation of subsequent crops by *Eldana saccharina*. Burning of crop residues was also reported to be effective against *Chilo* spp. and *Busseola fusca* as well [15]. Burning or composting old stalks before the onset of the rains is effective against *B. fusca* [13]. The management of the maize stalkborer *B. fusca* includes intercropping maize with non-hosts crops like cassava and cowpea or with a repellent plant such as silver leaf desmodium (*Desmodium uncinatum*) [33]. Others cultural practices such as destruction of alternatives host plants or ratoons, synchronized plantings, crop rotations, high cropping density, use of trap crops, good irrigation and good fertilization are good cultural tactics against the insect pests in general and stem borers in particular [27]. The use of trap crops or intercropping upland NERICA rice and maize have also been suggested as an effective method for controlling *M. separatella* in rice ecosystems in Nigeria [26]. Practices such as irrigation, planting density and dates of planting were all found to be effective as well important factors for consideration [17, 34]. According to [24], the management of the African rice gall midge take into account early and synchronized planting as rice fields planted early are less likely suffer serious damage than those planted late. Also destruction of alternative host plants such as rice ratoons, volunteers and *Oryza longistaminata* as well as the use of moderate levels of fertilizer (e.g. 60 kg/ha) prevent the build-up of AfRGM population. The same author highlighted the importance of plant spacing as close spacing provides a suitable micro-environment for the survival of the exposed life stages of AfRGM. Cleaning of the rice field especially the destruction of the wild rice are good cultural practice for managing the African striped rice borer, *C. zacconius*. The 'push-pull' method based on the intercropping of Desmodium with millet was report to effectively act as a repellent that 'pushes' the millet stem borer *C. ignefusalis* away from the millet [35]. The most useful advantage of the cultural practices is that they are compatible to all pest control measures. They represent an important component of the integrated pest management of the stems borers. The main disadvantage of the cultural practices is that they need to be continuous and collective process from field preparation to harvest. A good cultural practices field can get infested if the surrounding farmers do not apply same or no management option.

3.2 Varietal resistance

Plant resistance is the genetically inherited qualities that confer the plant ability to ward off or withstand pest attacks or recover from injury due to a pest [36, 37]. This method is the most farmer-friendly pest control option that can significantly reduce stem borer damage when supplemented with other options such as cultural or biological measures. It is most attractive as the use of insecticides is largely beyond the means of the small farmer. Considerable progress has been made in screening and breeding for host plant resistance to cereals' stem borers but only limited number of varieties have shown good level of resistance. Wiseman [38] showed that the resistant cultivar should be the base from which integrated pest management strategies arise. Rana *et al*. [39] reported that antibiotic property in sorghum plays more role in plant resistance to stem borer than ovipositional non-preference. Some plant biophysical characters such as stem hardness, leaf hairiness are important in plant resistance to stem borers. Sorghum varieties having these traits are rejected by the moths for oviposition. Pearl millet varieties such as Zongo was reported to be moderately resistant to *Coniesta ignefusalis* [40]. According to [41], hairiness of leaves and leaf sheaths were partly responsible for the differences in genotypic vulnerability to *C. ignefusalis*. Also they reported that plants with trichomes were not preferred by this pest for oviposition. [42] reported good level of resistance in the sweet sorghums BR 501, BR 504, and BR 505 to the sugarcane borer *Eldana saccharina*. [20], found good source of resistance to diopsids among upland NERCA varieties. Also, they reported that rice varieties having ability to produce new tillers to compensate the infested stems can tolerate the damage by diopsid. So far no improved rice variety was identified to be resistant to *O. oryzivora* attack but some tolerance was noted *in Oryza glaberrima* and also in some improved released rice varieties in Nigeria such as Cisadane and FARO 51 [24]. Despite limited achievement on varietal resistance to stem borer, this option remain a promising IPM component. Recent advances in biotechnology can increase the prospects of generating resistance materials and accelerate the transfer of gene for improving new genotypes.

3.3 Biological control measures

Biological control is the manipulation of natural enemies with the aim to maintain pest population below the economic injury level (EIL). Several organisms such as insects, fungus, virus and bacteria can be used as biocontrol agents [13, 43]. Insects based organisms acting as natural enemies are either predators (using the host as food) or parasitoids (laying their eggs in the host). Most of these insects belong to hymenoptera or diptera orders [27]. Biocontrol appears to be one of the most effective and environment friendly management option of stem borers. Indeed, stem borer's population and damage can be regulated by sustaining the action of natural enemies. This can be done through a good habitat management to favor the buildup of the population of natural enemies (spiders, wasps, ladybirds, etc.) or through mass rearing and field release of specific parasitoids to control target pest species. The success of the release of several parasitoids was reported in managing cereals stem borers in Africa. Two natural enemies of the maize stalkborer (*B. fusca*) are the larval parasitoids *Cotesia sesamiae* and *Bracon sesamiae* [31, 44]. Parasitoids such as *Tetrastichus atriclavus, Apanteles sesamiae,* and *Pediobius furvus* have been reported by [13] to be most important parasites of *B. fusca.* Similarly *Cotesia flavipes* and *Xanthopimpla stemmator* was reported to effectively control the spotted stem borer *C. partellus* [4]. The parasitoids *Cotesia sesamiae, Xanthopimpla stemmator, Trichogramma spp, etc. are cited as good biocontrol agent against the* pink Stem Borer *Sesamia calamistis* (Togola, unpublished data). The biological control of the cereal stem borers is mainly based on habitat management to sustain natural enemies including various parasitoids wasp. [45] found that that the contribution of egg parasitism is more important in controlling lepidopteran stem borers than parasitism of larvae and pupae. The African rice gall midge (AfRGM) is attacked by two parasitoids such as *Platygaster diplosisae* (Hymenoptera: Platygastridae) and *Aprostocetus procerae* (Hymenoptera: Eulophidae) that can decrease the population of the pest below the economic injury threshold in rice-production systems [25]. Several insects species such as *Cyrtorhinus viridis* (Heteroptera: Miridae), *Conocephalus longipennis* (Orthoptera: Tettigoniidae) and *Anaxipha longipennis* (Orthoptera: Gryllidae) are predators of AfRGM [24]. The effectiveness and sustainability of the biological control methods depends on the availability of the biocontrol agent at suitable density. Practices such as habitat management or avoiding the use of wide spectrum chemicals can contribute to increase the carry-over of population of natural enemies and maintain the pest population below a critical level. The main constraints of the biocontrol measures are the difficulty to find the specific biocontrol agents for targeted pest species, the complexity of the mass rearing and the complication to be explained by extension workers and to be implemented by farmers.

3.4 Chemical control

Chemical control, despite all the danger and environmental hazard, remains an important option to consider in situation where the pest population is already established. Also it can be used as IPM component to supplement varietal resistance or cultural practices. Chemical control can be achieved by applications of granules or dusts to the leaf whorl early in crop growth to kill early larval instars of *E. saccharina* [10]. Controlling *M separatella* using chemical insecticides is effective but not widely practiced because of the high costs involved [46]. As for *O. oryzivora*, chemical control can be envisaged in conditions of high infestation of rice field. In all cases, choice of selective systemic insecticides is needed to avoid adverse effects on non-target organisms and biodiversity.

3.5 Integrated pest management option

The individual control methods discussed above have their limitations and none often is sufficient to adequately control stem borer outbreaks. Hence, the integrated pest management (IPM), also known as integrated pest control (IPC) appears to be the most appropriate option for managing these pests. IPM requires the combination of several compatible and complementary practices with the aim to maintain pest populations below the economic injury level (EIL) while reducing the use of high hazardous pesticides and sustaining the action of natural enemies. Several studies have reported the success of IPM in the management of cereals' stem borers. [25] reported that varietal resistance/tolerance, cultural practices and biological control are important components of integrated management of rice stem borers. Similarly [24] found that the effective control of the African Rice Gall Midge relies on the combination of cultural practices, habitat management and moderate use of insecticide chemical. Kega [47] demonstrated that the use of resistant rice cultivars and entomopathogenic nematodes is a viable method to control *M. separatella.* Nwanze and Mueller [48] indicated that host plant resistance and cultural practices should be major components in the integrated management of sorghum stem borers. According to [49] an increase of yield can be obtained when sorghum varieties with tolerance or moderate resistance to stem borer are coupled with need-based application of pesticides. Youm et al. [41] suggested options such as early planting, destruction of crop residues and use pheromone bait traps for successful management of the millet stem borer *C. ignefusalis*. According to the conclusion from an international workshop organized by the International Institute for Semi-Arid Tropics, cultural methods and host plant resistance should be considered as the major components of the integrated management of cereals' stem borers [50]. However these practices need to be reinforced with other measures such as biological control and if necessary the use of selective systemic chemical . It is important to quote that integrated stem borer management is likely to be severely constrained by the limited capability of farmers to implement several options . For this reason it is highly important that the IPM takes into account the community farming systems and know-how.

4. Conclusion

The stem borers represent a group of insects of economic importance to cereal crops in Africa. Because of the nature of their attacks and the complexity of their biology, the success of the management options will depends on the integration of various strategies ranging from cultural practices to host plant resistance, biological control and moderate use of systemic chemical when necessary. The cultural practices and host plant resistance remains the major component of the IPM of cereal's stem borers. They can be reinforced by the biological and chemical control. The cultural practices involve farmers' engagement and cooperation. As for varietal resistance, more research action is needed to identify or develop varieties that tolerate the stem borers attack. Regarding the deployment of chemical and biocontrol options, more intensive action from extension service is need to increase the capacity of farmers so that they can engage appropriate action to limit yield losses in cereal and increase their incomes.

Author details

Abou Togola[1]*, Ousmane Boukar[1], Manuele Tamo[2] and Siva Chamarthi[1]

1 International Institute of Tropical Agriculture, Kano, Kano State, Nigeria

2 International Institute of Tropical Agriculture, Cotonou, Benin

*Address all correspondence to: a.togola@cgiar.org

References

[1] Bosque-Pérez NA, Mareck JH. Distribution and species composition of Lepidopterous maize borers in southern Nigeria. Bulletin of Entomological Research. 1990;**80**(4):363-368

[2] Schulthess F, Bosque-Pérez NA, Gounou S. Sampling lepidopterous pests on maize in West Africa. Bulletin of Entomological Research. 1991;**81**(3):297-301

[3] Annecke DP, Moran VC. Insects and Mites of Cultivated Plants in South Africa. Butterworth; 1982

[4] Akol AM, Chidege MY, Talwana HAL, Mauremootoo JR. *Chilo partellus* (Swinhoe, 1885)-Spotted Stemborer; 2011b

[5] Mohyuddin AI, Greathead DJ. An annoted list of the parasites of graminaceous stem borers in East Africa, with a discussion of their potential in biological control. Entomophaga. 1970;**15**(3):241-274

[6] Akinsola EA. Management of Chilo spp. in rice in Africa. International Journal of Tropical Insect Science. 1990;**11**(4-5):813-823

[7] Heinrichs EA, Barrion AT. Rice-feeding insects and selected natural enemies in West Africa biology. Ecology, Identification. 2004. GP Hettel Edited

[8] Dale D. Insect pests of the rice plant–Their biology and ecology. In: Heinrichs EA, editor. Biology and Management of Rice Insects. New Delhi: Wiley Eastern; 1994. pp. 363-485

[9] Sampson MA, Kumar R. Alternative host plants of sugarcane stem borers in southern Ghana. Insect Science and its Application. 1986;**7**:539-541

[10] Akol AM, Chidege MY, Talwana HAL, Mauremootoo JR. *Sesamia calamistis* Hampson, 1910 – African Pink Stalkborer;2011c

[11] Van Hamburg H. The Grain Sorghum Stalk Borer. Pretoria, South Africa: Farming in South Africa, Plant Protection Research Institute; 1979

[12] Tanzubil PB, Mensah GWK, McCaffery AR. Relationship between pheromone trap catches of Coniesta ignefusalisand larval infestation of a millet crop in Ghana. Tropical Science. 2003;**43**:63-69

[13] Harris KM. Lepidopterous stem borers of cereal crops in Nigeria. Bulletin of Entomological Research. 1962;**53**:139-171

[14] Reddy KS. Relative susceptibility and resistance of some sorghum lines to stem-borers in Western Kenya. International Journal of Tropical Insect Science. 1985;**6**(3):401-404

[15] Ingram WR. Biological control of graminaceous stem-borers and legume pod-borers. International Journal of Tropical Insect Science. 1983;**4**(1-2):205-209

[16] Brenière J. Importance des problèmes entomologiques dans le développement de la riziculture de l'Afrique de l'Ouest. L'Agronomie Tropicale. 1969;**24**:906-927

[17] Breniere J, Rodriguez H, Ranaisova H. Un ennemie riz a Madagascar, *Maliarpha separatella* Rag. ou borer blanc. L'agronomie Tropicale. 1962;**17**:223-302

[18] Khan ZR, Chiliswa P, Ampong-Nyarko K, Smart LE, Polaszek A, Wandera J, Mulaa MA. Utilisation of wild gramineous plants for management of cereal stemborers in Africa. International Journal of Tropical Insect Science. 1997;**17**(1):143-150

[19] Banwo OO. Management of major insect pests of rice in Tanzania. Plant Protection Science. 2002;**38**:108-113

[20] Togola A, Nwilene FE, Agbaka A, Degila F, Tolulope A, Chougourou D. Screening upland varieties of NERICA and its parents for resistance to stalk-eyed Fly, *Diopsis* sp. (Diptera, Diopsidae) in Benin. Journal of Applied Sciences. 2011a;**11**:145-150

[21] Togola A, Nwilene FE, Koné B, Chouqourou D. Sedentarisation of the population of Diopsid flies in Rice agro-systems in Benin. Tropicultura. 2011b;**29**(2):101-106

[22] Scheibelreiter G. The importance of *Diopsis tenuipes* Westwood as a pest of rice, based on a comparison of the egg-laying behaviour of *D. tenuipes* and *D. thoracica* Westwood. Ghana Journal of Agricultural Science. 1974;**7**:143-145

[23] Bennett J. New approaches to gall midge resistance in rice. International Rice Research Institute. 2004

[24] Nwilene FE, Nwanze KF, Okhidievbie O. African Rice Gall Midge: Biology, Ecology and Control. Field Guide and Technical Manual. Cotonou: Africa Rice Centre; 2006

[25] Nwilene FE, Togola A, Agunbiade TA, Ogah EO, Ukwungwu MN, Hamadoun A, et al. Parasitoid biodiversity conservation for sustainable management of the African rice gall midge, Orseolia oryzivora (Diptera: Cecidomyiidae) in lowland rice. Biocontrol Science and Technology. 2008;**18**(10):1075-1081

[26] Nwilene FE, Nacro S, Tamò M, Menozzi P, Heinrichs EA, Hamadoun A, et al. 18 Managing Insect Pests of Rice in Africa. Realizing Africa's Rice Promise2013. p. 229

[27] Kumar R. Insect Pest Control. 41, Bedford Square, London WCl 3DQ: Editions Edward Arnold Ltd. KARTHALA Editions; 1991

[28] Ajayi O. Stem borers of sorghum in West Africa with emphasis on Nigeria. In: International Workshop on Sorghum Stem Borers. Vol. 17. 1987, November. p. 20

[29] Sarwar M. Effects of zinc fertilizer application on the incidence of rice stem borers (Scirpophaga species) (Lepidoptera: Pyralidae) in rice (*Oryza sativa* L.) crop. Journal of Cereals and Oilseeds. 2011;**2**(1):61-65

[30] Sarwar M. Effects of potassium fertilization on population build up of rice stem borers (lepidopteron pests) and rice (*Oryza sativa* L.) yield. Journal of Cereals and Oilseeds. 2012;**3**(1):6-9

[31] Kfir R. Parasitoids of the African stem borer, Busseola fusca (Lepidoptera: Noctuidae), in South Africa. Bulletin of Entomological Research. 1995;**85**(3):369-377

[32] Sampson MA, Kumar R. Population dynamics of the stem-borer complex on sugar-cane in southern Ghana. International Journal of Tropical Insect Science. 1983;**4**(1-2):25-32

[33] Akol AM, Chidege MY, Talwana HAL, Mauremootoo JR. *Busseola fusca* (Fuller, 1901)–African Maize Stalkborer;2011a

[34] Ukwungwu MN. Planting time and stem borer incidence in Badeggi, Nigeria. International Rice Research Newsletter. 1984;**9**(1):22

[35] Chernoh E. Millet Stem Borer Coniesta ignefusalis. Technical Sheet. CABI; 2014 http://africasoilhealth.cabi.org/wpcms/wp-content/uploads/2015/02/48-cereals-millet-stem-borer1.pdf

[36] Painter RH. Insect resistance in crop plants. In: The Macmillan Company. New York; 1951

[37] Togola A, Boukar O, Belko N, Chamarthi SK, Fatokun C, Tamo M, et al. Host plant resistance to insect pests of cowpea (*k* L. Walp.): Achievements and future prospects. Euphytica. 2017;**213**(11):239

[38] Wiseman BR. Host plant resistance in crop protection in the 21st century. In: 11. International Congress of Plant Protection, Manila (Philippines). 1987, 5-9 Oct 1987

[39] Rana BS, Singh BU, Rao NGP. Breeding for Shoot Fly and Stem Borer Resistance in Sorghum. Pages 347-360 in Proceedings of the International Sorghum Entomology Workshop, 15-21 Jul. 1984, College Station, Texas, USA. Patancheru, A.P. 502324. India: International Crops Research Institute for the Semi-Arid Tropics; 1985

[40] N'Doye M. Syntheses de quelques resultats sur les insectes foreurs des mille et sorgho au Senegal. Bambey, Senegal: Center de Recherches Agronomiques; 1977. p. 15

[41] Youm O, Harris KM, Nwanze KF. Coniesta Ignefusalis (Hampson), the Millet Stem Borer: A Handbook of Information1996. p. 53

[42] Reyes R. Sorghum stem borers in Central and South America. In: Proceedings of the International Workshop on Sorghum Stem Borers. ICRISAT Center; 1987, November. pp. 49-60

[43] Gahukar RT. Biological control of insect pests of sorghum and pearl millet in West Africa. In: Biological Control of Pests: Its Potential in West Africa. Dakar, Senegal; 1981, 9-13 February, 1981

[44] Kfir R. Natural control of the cereal stemborers *Busseola fusca* and *Chilo partellus* in South Africa. International Journal of Tropical Insect Science. 1997;**17**(1):61-67

[45] Mathez FC. *Chilo partellus* Swinh., *C. orichalcociliella* stand (Lep: Crambidae) and *Sesamia calamistis* Hmps. (Lep: Noctuidae) on maize in the Coast Province, Kenya. Mitteilungen der Schweizerischen Entomologi-schen Gesellschaft. 1972;**45**:267-289

[46] Akinsola EA, Agyen-Sampong M. The ecology, nbionomics and control of rice stem-borers in West Africa. Insect Science and its Application. 1984;**5**:69-77

[47] Kega VM, Kasina M, Olubayo F, Nderitu JH. Management of Maliarpha separatella rag using effective entomopathogenic nematodes and resistant rice cultivars. Journal of Entomology. 2013;**10**(2):103-109

[48] Nwanze KF, Mueller RAE. Management options for sorghum stem borers for farmers in the semi-arid tropics. In: Proceedings of the International Workshop on Sorghum Stem Borers. ICRISAT Center; 1989. pp. 105-116

[49] Kalode MB, Bentur JS, Srinivasan TE. Screening and breeding rice for stem borer resistance. In: Proceedings of the International Workshop on Sorghum Stem Borers. 1987, November. pp. 17-20

[50] ICRISAT (International Crops Research Institute for Semi-Arid Tropics). International workshop on sorghum stem borers, 17-20 Nov, 1987, ICRISAT Center, India Patancheru. 1989. AP. 502324 India: ICRISAT

Section 3

Acarology

Chapter 5

Prologue: Scientific and Societal Importance of Mites and Acarology From the Viewpoint of International Publication

Levente Hufnagel, Ferenc Mics and Réka Homoródi

1. Introduction

Mites are arthropods typically (and mostly) with a size of 0.08–1 mm; however, exceptionally they might have individuals even with a size of 3 cm among the ticks (Ixodida) when full of blood. Nowadays, they are rather considered to be a morphological or life-form group than an evolutional zoosystematic concept. By comparing the body size, the size range of terrestrial soil mesofauna is 0.1–2 mm and the size range of insects is 0.2–300 mm, whereas in water ecosystems, microplankton has a range of 0.02–0.2 mm and mesoplankton has a range of 0.2–20 mm.

It is doubtless that all of them belong to the class Arachnida of the subphylum Chelicerata; however, the monophyletic origin of the subclass Acari is in question. Probably their groups Anactinotrichida (Parasitiformes) and Actinotrichida (=Acariformes) have to be considered as separate subclasses. Chelicerates were the first multicellular animals on land (in the Silurian), and their sea life-forms are known from the Cambrian Period on; however, their ancestors might have been present in the Ediacaran fauna as well, although their relationship with the Trilobites is not sufficiently explained. Back to the mites, Actinotrichida is known from the Devonian period. Several of their fossils cannot be distinguished morphologically from recent genera. Their relationship with the subclass Solifugae has been suggested. However, their group Anactinotrichida is known only from the Cretaceous Period on; it is not impossible that they form a monophyletic group with the subclass Ricinulei (which do not belong to the mites); however, their relationship with the group Pseudoscorpiones has also been suggested. They are mostly oviparous; however, viviparism also occurs, and even mature offspring may be born (e.g. Pyemotidae). Postembryonal development of the arachnids is usually epimorphosis; however, hemi-metamorphosis can be observed among the mites (with separate larval form and various nymphal forms; even the number of legs can vary from two to three pairs); this phenomenon also shows similarity with the subclass Ricinulei. Sometimes the postembryonal development is very complex, that of the male and the female can be different, which allows them to have different ecological functions. The phylogeny of Chelicerates is summarised by [1].

IntechOpen

2. The scientific importance of mites

Mites form the most prevalent and diverse group of the subphylum Chelicerata. They are basically and mostly terrestrial animals; however, they also have freshwater and sea species secondarily. Multicellular animals which returned from the land to permanent sea life for the first time in the history of the Earth (in the Permian) belong here (Trombidiformes: Halacaridae). As for feeding, they have herbivorous, fungivorous, deposit feeder, predator and parasite species as well, which is a unique (and evolutionary secondary) feature among the mostly predator arachnids. Although their active motion is not efficient due to their small body, anemochorous, hydrochorous, biochorous and anthropocorous methods of dispersal are also known; thus, many of their species are cosmopolitans. Many of their ancient species live in tropical rain forests, moss forests and bogs, whereas in the temperate and arctic zones as well as in the (even tropical) mountains, their younger, derived forms are frequent. In South America, in the higher regions of the Pacific Mountain Range, Holarctic fauna can extend as far as South Chile, whereas at a lower height above sea level, the neotropical fauna becomes dominant in the southern part of Central America. At the same time, in the mountains of Papua New Guinea, in high elevations, their Antarctic species are present [2]. They are abundant in all biogeographical regions and all biomes, from the tundra through semi-deserts to the rain forests, from the Antarctica to the Arctic. They are extremely resistant to both physical and chemical impacts. Several of their species can tolerate dehydration well and others are able to live in sea water permanently (deep in the zone of eternal darkness), whereas others can tolerate the total freezing of the substrate surrounding them. The permanent darkness and the low oxygen level do not cause problems either. In most of their species resistance had developed to the special acaricides as well [3–5]. In case of a record-breaking mass extinction due to the destruction of the environment by mankind, they will have representatives among the survivors, since many of their groups had survived the most devastating Permian-Triassic extinction to date almost without any change.

The number of their taxa is very uncertain, the number of their species described so far exceeds 80,000, the number of their genera exceeds 3000 and that of their families exceeds 800. However, these numbers do not mean much. On the one hand, based on the pace of the species descriptions, the majority of species living on the Earth (even 90–95%) is still unknown for the science; on the other hand, their morphologically described species may prove to be several different species based on molecular studies. At the same time, it is doubtless that the majority of species was described by museologists who had the interest in increasing the number of species names in favour of their scientific career; thus it cannot be precluded that up to three-fourths of the described species has to be synonymised in the future. Their sometimes surprisingly significant sexual dimorphism may also play a role in this. Finally, uncertainty is increased by the fact that permanently parthenogenetic, asexual populations also occur among the mites. In this case, the definition of species itself becomes uninterpretable, since even each of their individuals can be considered as a separate species due to the existing gene flow barrier and the individual genetic differences which can certainly be easily demonstrated. However, it is doubtless that we face huge morphological, ecological and molecular diversity.

The research of mites is of great importance due to their quantity and role in the cycle of materials, since there can be tens of thousands of individuals belonging to up to 300–400 species in a handful of tropical soil, moss, bark, hanging soil, bromeliad funnel or tussock. There is no flat, city park, agricultural area, rubbish heap, tumbledown building or any other habitats where they are not abundant. They can be determined in the decomposition in the soils and especially in the inimitably waste-free cycle of materials in the floodplain tropical forests.

3. The social importance of mites

The role of mites (especially that of oribatid mites but also that of prostigmatic, mesostigmatic and astigmatic mites) in the cycle of materials in soils is not fully explained yet; however, their importance in agriculture and forestry is obvious (especially in the tropics). Without them, the release of plant nutrients from corpuscular organic materials would be much slower, and efficient agricultural production would be impossible in many cases. However, some phytophagous groups (e.g. Tetranychoidea, Tarsonemidae, Eriophyidae and Penthaleidae) can be significant pests in horticulture, agriculture and forestry. Research on the biological pest control of *Tetranychus urticae* is of special importance [6, 7]. Besides, they can be parasites and disease vectors (e.g. Varroa spp.) of bee species which have a significant role in pollination. On the contrary, other, mostly predator, mite groups contribute to biological pest management and ecological farming (Phytoseiidae).

The minority of mites is of public health or animal health importance as parasites (e.g. Sarcoptidae, Psoroptidae, Knemidokoptidae, Demodicidae, Cheyletidae), as vectors of pathogens (e.g. Ixodidae, Argasidae, Dermanyssidae) or due to the allergenic impact of their faeces (e.g. Pyroglyphidae). Forensic acarology is also an interesting field of application, where they can be used as bioindicators. As indicators, they are very promising in environmental and nature protection as well [8–15].

4. Volume and composition of the acarological literature according to the data in the large publication databases

Top quality but minor part of scientific papers is published in English-language journals with impact factor (these articles are available in the Web of Science Core Collection database dating back to 1975). Collecting the older scientific literature is more difficult; however, the Zoological Records database, which is part of the wider collection range of the Web of Science, is helpful in this case. A less selected but still internationally noted part of the scientific journals can be found in the Scopus database. However, these are still a fraction of the large amount of articles which were and are virtually published. Google Scholar provides a small insight into this, and it collects all scientific articles available on the Internet; however, its scientific criteria are much weaker. There are a large number of scientific journals available only in print and in national languages all over the world which are published by scientific research institutes, universities and museums but are not included in the above-mentioned large databases; the quality of their review and the level of their editorial board is often uncertain.

In order to be able to get a picture of the volume of the acarological research, an analysis was made on the results of the available important databases and information sources using some appropriate keywords (**Table 1**). We chose the Latin "Acari" and the English "mites" words as keywords, and the Hungarian "atkák" word was

Keyword	WoS Core Collection	Scopus	WoS all databases	Google Scholar	MATARKA	Google internet
"Acari"	19,805	26,903	50,975	275,000	572	5,030,000
"Mites"	36,605	43,186	49,128	650,000	144	30,300,000
"Atkák"	0	2	16	257	90	86,200

Table 1.
Number of results obtained with keywords related to acarology in different databases (status: 22 September 2018).

added as well in order to obtain some information also on the articles published in minor languages. Accordingly, besides the international databases, the national database of journals published in Hungary (MATARKA) was added as well. The number of the Hungarian results may help with estimating the potential volume of other hidden publications as a special indicator, considering that among the 243 countries (195 independent states) of the world, Hungary ranks 92nd regarding the size of the population, and it amounts to 0.13% of the world's population.

Finally, the table was completed with the number of results from the Google browser, since these data and proportions help with evaluating the obtained results.

The number of results in the scientific databases shows that the number of scientific articles amounts to 0.1–5% of the general frequency on the Internet; this is similar in the case of the Latin, English and Hungarian words. Furthermore, it can be stated that the number of top quality articles (WoS CC) amounts to only 5–7% of all scientific results on the Internet (Google Scholar).

The explosive increase in the intensity of scientific research is well characterised by the temporal distribution of the number of results regarding English and Latin keywords (**Table 2**).

On the graphs it can be seen that the increase in the research intensity is exponential, and the fall back in the last years is an artefact, which can be explained by the time needed for data to be included in the databases. At the same time, in the case of the Latin term, a real stagnating period can be recognised in the twenty-first century, which is not shown by the English keyword. This certainly refers to a break in the traditional taxonomic, faunistic and museological approaches compared to other (e.g. molecular and quantitative ecological or applied) studies.

Acari Scopus: 26,903 docs (September 22, 2018)	Mites Scopus: 43,186 docs (September 22, 2018)
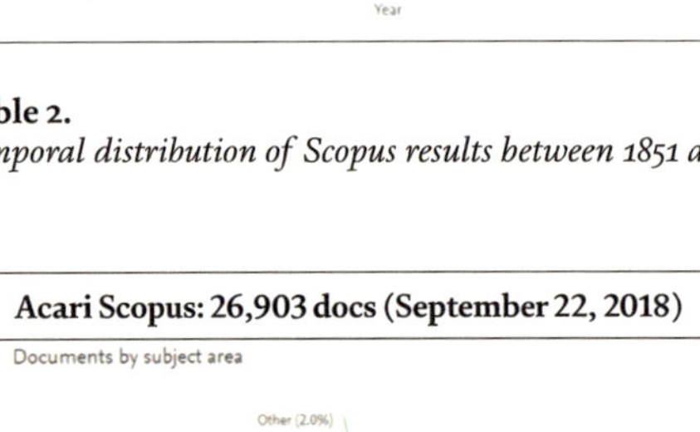	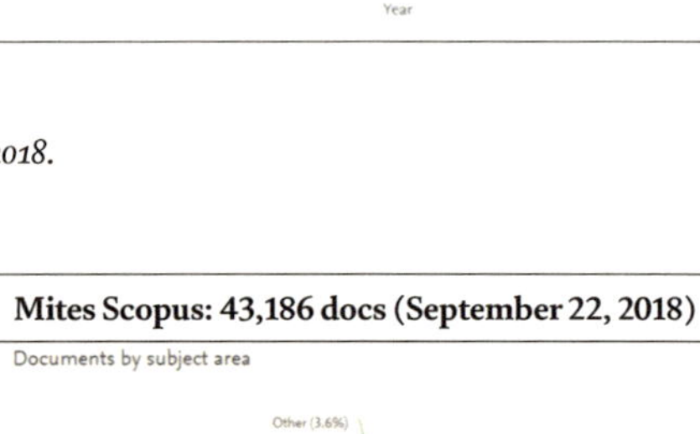

Table 2.
Temporal distribution of Scopus results between 1851 and 2018.

Acari Scopus: 26,903 docs (September 22, 2018)	Mites Scopus: 43,186 docs (September 22, 2018)
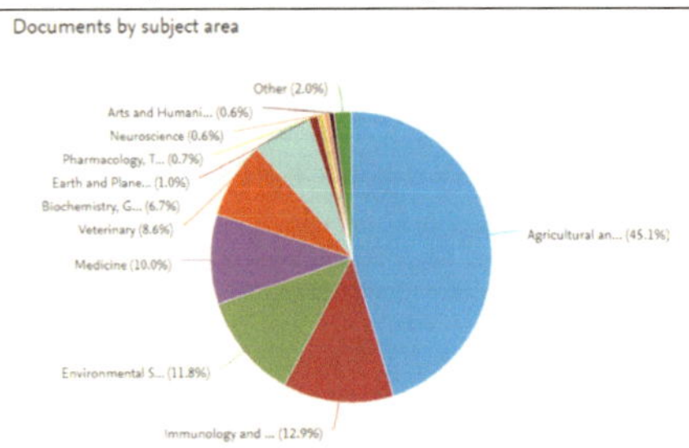	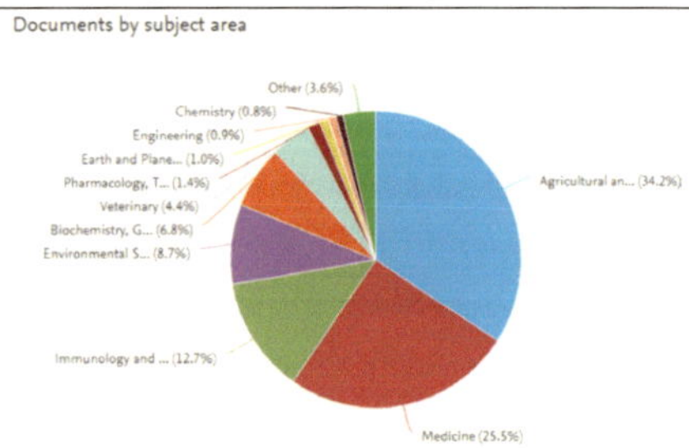

Table 3.
Distribution of Scopus results based on subject area between 1851 and 2018.

Acari Scopus: 26,903 docs (September 22, 2018)	Mites Scopus: 43,186 docs (September 22, 2018)
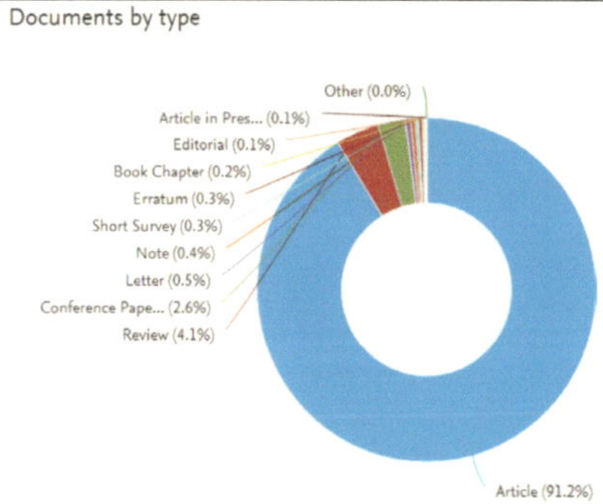	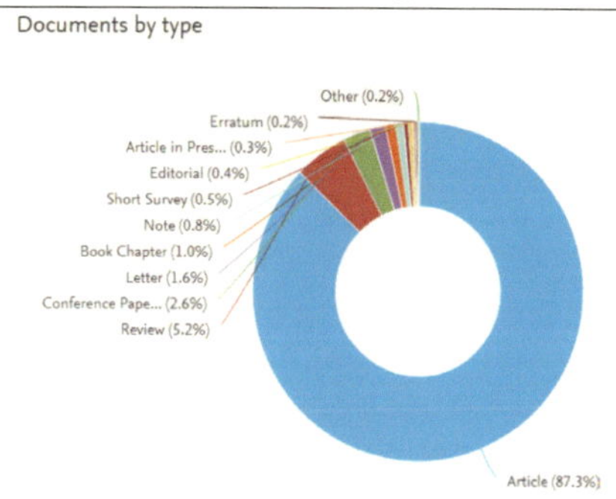

Table 4.
Distribution of Scopus results based on type between 1851 and 2018.

Acari Scopus: 26,903 docs (September 22, 2018)	Mites Scopus: 43,186 docs (September 22, 2018)
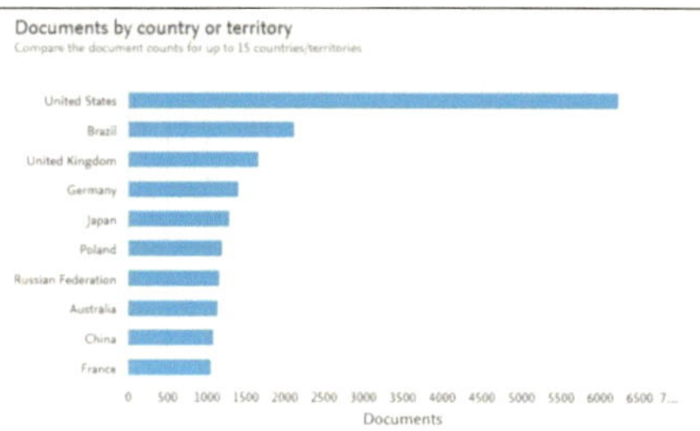	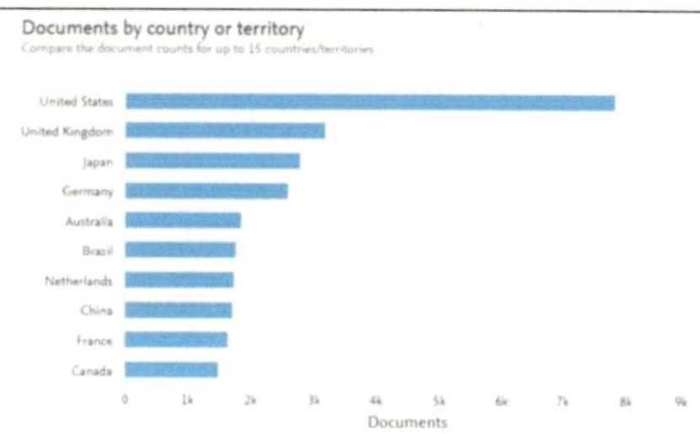

Table 5.
Distribution of Scopus results based on countries between 1851 and 2018.

Studying the distribution of the articles based on the subject area (**Table 3**), one can see that more than three-fourths of them belongs to applied fields of research (agricultural science, human health, environmental protection).

Studying the distribution of publications based on the type (**Table 4**), the high dominance of primary articles is clear, which refers to the low synthesis level of the subject area. This can have serious disadvantages regarding the application of new knowledge and the knowledge transfer.

Based on the geographical distribution by country (**Table 5**), the results show the dominance of the United States in this subject area so far. The contribution of Brazil, the United Kingdom, Germany and Japan seems to be significant as well.

5. Acarological research among elite publications considered to be an outstanding scientific breakthrough

It is worth studying in which extent acarological research contributed to outstanding scientific breakthroughs in the last four decades (from 1975 to February 2019). This study was based on the articles of the journals *Nature* and *Science*, two absolutely elite organs of scientific research, using the keywords "Acari" and "mites". A considerable article was published in *Nature* in 2013, whereas in *Science* in 2019 for the last time. The results can be seen in **Table 6**. The results obtained with the overview of the previous publications of greater volume are nuanced by 25% of the publications which were results using the keyword; however, by overviewing 55 articles, it can be stated that those do not have real acarological relevance.

	Nature	Science
Allergy and health	5	5
Agriculture and plant-Acari interactions	6	8
Ecology and evolutionary biology	8	9
Mentions without real acarological relevance	6	8
Total	25	30

Table 6.
Articles related to acarology in Nature and Science.

Furthermore, it can be stated that 42% of the elite publications with real acarological relevance have solely basic research (ecological and evolutionary biological) motivation, and 34% of them have agricultural science and 24% human health motivation. Among the articles published in *Nature*, the publication by [16] on the genome of *Tetranychus* urticae was cited on an outstanding number of times.

Another approach to elite publications is studying the most cited acarological articles in the WoS Core Collection database. Among the 60 most cited articles, 32 are related to allergy or human health, 19 are related to agriculture (11 to pest management and 8 to apiary), 5 are related to ecological basic research and 4 of them are without real acarological relevance. Concerning acarology, the following publications had an especially outstanding citation impact: [17] on the biology of *Varroa destructor*, [18] on the global patterns of soil communities, [16] on the genome of *Tetranychus urticae*, [19] on *Rhipicephalus sanguineus* related to veterinary parasitology as well as [20] on the European distribution of *Ixodes ricinus*.

6. Forefront of acarological research

The repository of the top quality and internationally recognised publications is the WoS Core Collection database. In WoS Core Collection, there are 2763 results (published between 1 January 2018 and 15 February 2019) with the keyword "Acari" or "mites" in their topic. Among these, 64% are related to human health, 35% have classical zoological approach, only 4% have ecological approach, also 4% have agricultural approach and only 5% have multidisciplinary approach (articles may belong to several categories; thus, these percentages cannot be added up).

This proportion is striking, since it shows that a significant part of the research on such an important group is not related to ecological, sustainability, agricultural and multidisciplinary research connected with the global ecological crisis of our time and the sixth mass extinction in the geological present, but to the diseases of a single species (human). What is more, it is not related to diseases considered to be the main causes of death, but mostly to allergy, whose main agents are not mites.

Among 2763 publications, 2682 were written in English, the second most important language is Spanish with 20 articles, then German comes with 16 articles and more than one article was written in Turkish, Japanese, Portuguese, Russian, French, Polish and Korean. (The above-mentioned Hungarian language is represented by one article.) This is natural, since the international language of science is basically English, and this database contains only articles of elite journals with impact factor conforming to the highest scientific criteria.

However, the ranking of research by countries is much more important than by languages, where behind the United States, China comes in second, preceding

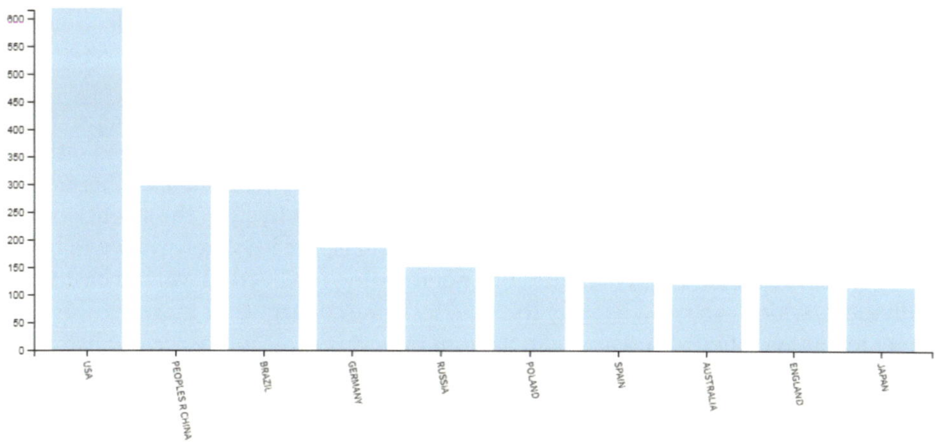

Figure 1.
Frequency distribution of acarological papers based on countries (WoS Core Collection).

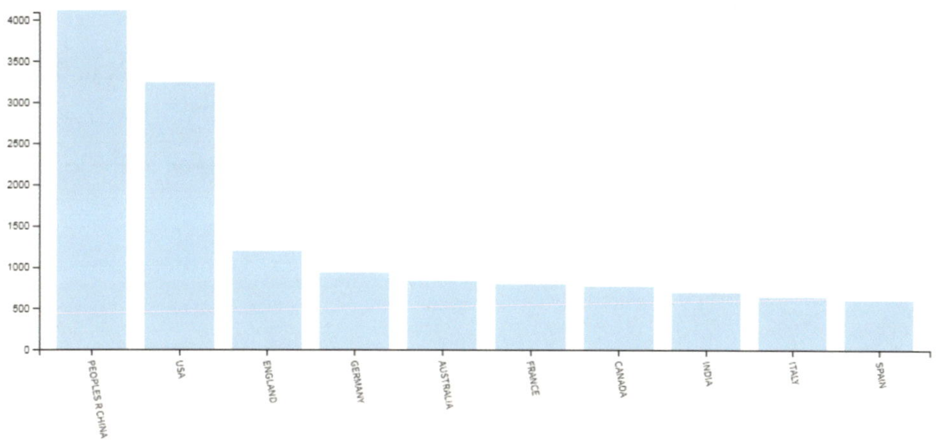

Figure 2.
Frequency distribution of environmental papers based on countries (WoS Core Collection).

Brazil and significantly outpacing the United Kingdom, which rank well before it in the summary relating to all time (**Figure 1**). The rise in Chinese scientific research is rather spectacular in other subject areas as well; this must set an example for all countries of the world.

If we do not search for mites but for the keyword combination "global problems or ecology or environment" in the same database for the same period, among the found 15,431 sources, the majority is published by Chinese authors (**Figure 2**). It would be worth studying the Chinese research funding and educational system; however, this is not the subject of this book.

The studied 2763 acarological publications were cited by other works 1794 times (1084 without self-citation) until the deadline of this study (16 February 2019), the h-index equals to 10 and the average number of citations is 0.65. There are 13 publications which are already cited more than 10 times. Among them, five are related to allergology and four to agricultural plant protection. None of the 10 articles was published in a specifically acarological journal; however, it is gladsome that two reviews can be found among them, which can improve the low synthesis level of the subject area.

7. Future perspectives, main directions and tasks of acarological research

In the front-rank scientific research, the main future research directions and tasks cannot be determined, and the real perspectives cannot be overviewed either, since these are results of the individual creativity of researchers and are inherently unpredictable. According to an article published in *Nature* by [21], the scientists of the University of Chicago, the larger a research team is, the less likely they will have really creative and innovative results. The outstanding intellectual work is basically individual, which can be efficiently fostered by some colleagues at most. However, larger teams and aligned research can have an important role in the accomplishment of existing ideas, setting up large databases and supporting future brilliant research. Concerning these, the following main tasks can be highlighted:

1. Taxonomical and biogeographical exploration of mite biodiversity

2. Quantitative coenological exploration of communities in intact ecosystems concerning mites as well

3. Preciser exploration of production biological roles of mites, especially in floodplain tropical forests and other natural and seminatural habitats

4. Exploration of phylogenetic relationships of mites up to the level of the known genera

Author details

Levente Hufnagel[1,2]*, Ferenc Mics[1] and Réka Homoródi[2]

1 Faculty of Agricultural and Environmental Science, Laboratory of Biometrics and Quantitative Ecology, Institute of Crop Production, Szent István University, Gödöllő, Hungary

2 ALÖKI Applied Ecological Research and Forensic Institute Ltd., Budapest, Hungary

*Address all correspondence to: leventehufnagel@gmail.com

References

[1] Dunlop JA. Geological history and phylogeny of Chelicerata. Arthropod Structure and Development. 2010;**39**(2-3):124-142. DOI: 10.1016/j.asd.2010.01.003

[2] Balogh P, Gergócs V, Farkas E, Farkas P, Kocsis M, Hufnagel L. Oribatid assemblies of tropical high mountains on some points of the "Gondwana-bridge"—A case study (methodological possibilities of coenological indication based on oribatid mites №. 1.). Applied Ecology And Environmental Research. 2008;**6**(3):127-158. Available from: http://epa.oszk.hu/02500/02583/00012/pdf/EPA02583_applied_ecology_2008_03_127-158.pdf

[3] Adesanya AW, Franco E, Walsh DB, et al. Phenotypic and genotypic plasticity of acaricide resistance in populations of *Tetranychus urticae* (Acari: Tetranychidae) on peppermint and silage corn in the Pacific Northwest. Journal of Economic Entomology. 2018;**111**(6):2831-2843

[4] Xin T, Li X, Yin J, et al. Three superoxide dismutase genes from *Tetranychus cinnabarinus* (Boisduval) involved in the responses to temperature and acaricide stresses. Systematic and Applied Acarology. 2019;**24**(1):16-32

[5] Wei P, Li J, Liu X, et al. Functional analysis of four upregulated carboxylesterase genes associated with fenpropathrin resistancein *Tetranychus cinnabarinus* (Boisduval). Pest Management Science. 2019;**75**(1):252-261

[6] Yeşilayer A. The repellency effects of three plant essential oils against the two-spotted spider mite *Tetranychus urticae*. Applied Ecology and Environmental Research. 2018;**16**(5):6001-6006. Available from: http://epa.oszk.hu/02500/02583/00055/pdf/EPA02583_applied_ecology_2018_05_60016006.pdf

[7] Li XY, Munir S, Cui WY, He PJ, Yang J, He PF, et al. Genome sequence of *Bacillus velezensis* W1, a strain with strong acaricidal activity against two-spotted spider mite (*Tetranychus urticae*). Applied Ecology and Environmental Research. 2019;**17**(2):2689-2699. Available from: http://www.aloki.hu/pdf/1702_26892699.pdf

[8] Gergócs V, Hufnagel L. Application of oribatid mites as indicators. Applied Ecology and Environmental Research. 2009;**7**(1):79-98. Available from: http://www.aloki.hu/pdf/0701_079098.pdf

[9] Hufnagel L, Gergócs V, Garamvölgyi Á, Homoródi R. Seasonal change of oribatid mite communities (Acari, Oribatida) in three different types of microhabitats in an oak forest. Applied Ecology and Environmental Research. 2011;**9**(2):181-195. Avilable from: http://unipub.lib.uni-corvinus.hu/1476/

[10] Gergócs V, Hufnagel L. Oribatid mites (Acari: Oribatida) in microcosms—A review. Applied Ecology and Environmental Research. 2011;**9**(4):355-368. Available from: http://epa.oszk.hu/02500/02583/00025/pdf/EPA02583_applied_ecology_2011_04_355-368.pdf

[11] Gergócs V, Hufnagel L. A new method to evaluate habitat status based on the use of data on Oribatid mites (Acari: Oribatida). In: Salampasis M, Matopoulos A, editors. Proceedings of the International Conference on Information and Communication Technologies for Sustainable Agri Production and Environment. Skiathos: HAICTA; 2011. pp. 63-75. Available from: http://ceur-ws.org/Vol-1152/paper6.pdf

[12] Gergócs V, Hufnagel L. Comparing the natural variation of oribatid mite communities with their changes associated with anthropogenic

disturbance. Environmental Monitoring and Assessment. 2017;**189**(4):203

[13] Gergócs V, Hufnagel L. Global pattern of oribatid mites (Acari: Oribatida) revealed by fractions of beta diversity and multivariate analysis. International Journal of Acarology. 2015;**41**(7):574-583

[14] Gergócs V, Rétháti G, Hufnagel L. Litter quality indirectly influences community composition, reproductive mode and trophic structure of oribatid mite communities: A microcosm experiment. Experimental and Applied Acarology. 2015;**67**(3):335-356

[15] Gergócs V, Homoródi R, Hufnagel L. Genus lists of oribatid mites—A unique perspective of climate change indication in research. In: Lameed GA, editor. Biodiversity Conservation and Utilization in a Diverse World. Rijeka: IntechOpen; 2012. pp. 175-208

[16] Grbic M, Van Leeuwen T, Clark RM, et al. The genome of *Tetranychus urticae* reveals herbivorous pest adaptations. Nature. 2011;**479**(7374):487-492

[17] Rosenkranz P, Aumeier P, Ziegelmann B. Biology and control of Varroa destructor. Journal of Invertebrate Pathology. 2010;**103**:S96-S119

[18] Fierer N, Strickland MS, Liptzin D, et al. Global patterns in belowground communities. Ecology Letters. 2009;**12**(11):1238-1249

[19] Dantas-Torres F. The brown dog tick, Rhipicephalus sanguineus (Latreille, 1806) (Acari : Ixodidae): From taxonomy to control. Veterinary Parasitology. 2008;**152**(3-4):173-185

[20] Medlock JM, Hansford KM, Bormane A, et al. Driving forces for changes in geographical distribution of *Ixodes ricinus* ticks in Europe. Parasites and Vectors. 2013;**6**(1)

[21] Wu L, Wang D, Evans JA. Large teams develop and small teams disrupt science and technology. Nature. 2019. DOI: 10.1038/s41586-019-0941-9

Chapter 6

Ecological Spotlights on Mites (Acari) in Norwegian Conifer Forests: A Review

Sigmund Hågvar

Abstract

Long-term studies on mites in Norwegian coniferous forests are summarized. In podzol soil with raw humus, mite densities could pass 1 million per m^2, with 48 species of Oribatida and 12 species of Mesostigmata. Field and laboratory experiments with liming and artificial acid rain showed that soil pH affected the structure of the mite community. Certain species of mites and springtails typical for acid soils did, however, show preference for a higher pH in monoculture. We hypothesized that competition could be a strong regulating factor in microarthropod communities. Several oribatid species were flexible regarding soil type, vegetation, substrate, and decomposition stage. The genus *Carabodes* showed examples on specialists: two species were grazers on *Cladonia* lichens in dry pine forests, while three were decomposers in dead polypore fungi. Another three oribatid species from different genera were unique in excavating spruce needles, producing slowly decomposing excrements, and probably contributing to stable, carbon-storing humus. In microcosms, predatory Gamasina mites were seen to regulate microarthropod numbers. Mites were able to adjust both their vertical and horizontal distribution in soil according to environmental change. A local and temporary burst of fungal activity could rapidly attract opportunistic fungal feeders. Several mites were active under snow, often feeding. Some even penetrated into the snow layer.

Keywords: Acari, coniferous forest, ecology, mites, Norway, Oribatida, review, soil pH

1. Introduction

Nowhere else, in nature, organisms are so densely packed as in soil. Combined with a huge number of species, "biodiversity in the dark" has fascinated biologists for long. In concert, soil organisms play a key role in terrestrial ecosystems, being of fundamental importance for plant growth, sustainable crop production, and biogeochemical cycling of nutrients. At the same time, soil biodiversity is vulnerable to human disturbance of different kinds. There is a critical need for understanding soil processes, how soil organisms respond to global change, and to take measures for long-term protection of soil biodiversity [1].

Mites (Acari) represent one of the species rich and abundant soil animal groups. Oribatid mites alone cover five feeding guilds, including the ability to digest chitin [2], and they represent four trophic levels in the decomposition process [3].

Another mite group, Mesostigmata, contains a multitude of predator species which control other microarthropod populations, both in the soil and in vegetation [4, 5]. Forest habitats, especially old forests with a well-developed litter layer, tend to have a high mite density, often with a species-rich fauna of oribatids [6–8].

Norwegian coniferous forests represent the western outpost of the Eurasian taiga. This giant forest belt, which is dominated by Norway spruce (*Picea abies* (L.) H. Karst.) and Scots pine (*Pinus silvestris* L.), typically contains a well-developed raw humus layer which represents a considerable global carbon storage. The slowly decomposing needles, cones, and other litter items in the forest floor create a fungus rich and sometimes deep, humus world, in which several mite groups thrive, including many oribatid species.

The present review is a synthesis of mite studies in coniferous forest soils of Southern Norway, published over a 40-year period [9–26]. In the 1970s, extensive studies on soil microarthropods were initiated as part of a large project, "Effects of acid precipitation on forest and soil," and certain subjects were followed up long after the project was ended. In addition to summing up field and laboratory experiments with liming and artificial acid rain, spotlights will be given on the following topics: density and species numbers of mites, their horizontal and vertical distribution, effects of different pH, vegetation types, soils and substrates, succession in the mite fauna during decomposition, whether mites can influence the humification process, how species within one genus may differ in habitat use, an experiment on the predatory effect of Gamasina mites, and mite activity beneath and within snow.

2. Material and methods

2.1 Study areas

2.1.1 Main study area: Nordmoen

This was a spruce forest with *Vaccinium myrtillus* L. vegetation, situated on a flat plain of glaciofluvial sandy deposits, about 45 km N of Oslo. On clearcut areas, *Deschampsia flexuosa* (L.) dominated. The soil was a stone-free iron podzol with a 3 cm thick organic layer and a correspondingly bleached layer below. Experiments with artificial acidification and liming and decomposition experiments with litter bags were performed here, partly in a young spruce stand, and partly on a clearcut area [14].

2.1.2 Two study areas covering the range of coniferous forest types: Ås (A) and Skrukkelia (B)

Two study areas were chosen for soil sampling in natural forest, each area with a gradient in vegetation types from the poorest pine forest to the richest spruce forest [15, 27]. Area A near Ås, about 30 km south of Oslo, had a cover of marine sediments. In area B in Skrukkelia, NW of lake Hurdalssjøen and about 60 km north of Oslo, the soil was mainly morainic deposits. In both study areas, spruce forest with *Vaccinium myrtillus* dominated. Listed after increasing soil fertility based on plant associations, the vegetation types were short named as follows:

1. *Cladonia* sp.: pine forest on iron podzol soil, with a dense cover of *Cladonia* lichens. Due to a thin soil layer, conditions were dry, and trees grew slowly (**Figure 1**).

2. *Calluna vulgaris:* pine forest with less *Cladonia,* and a field layer dominated by *Calluna vulgaris* (L.) Hull. The soil was shallow peat in area A and iron podzol in area B.

3. *Vaccinium sp.:* pine forest on iron podzol soil, with a dense cover of *Vaccinium myrtillus* or *Vaccinium vitis-idaea* L., but also containing some *Cladonia* lichens.

4. *Vaccinium myrtillus:* spruce forest with *Vaccinium myrtillus*. Brown earth-like soil in area A and iron podzol in area B.

5. *Small ferns:* spruce forest with small ferns, *Dryopteris phegopteris* (L.) C. Chr. and *Dryopteris linnaeana* C. Chr. Brown earth in area A and iron podzol in area B.

6. *Small herbs:* spruce forest on brown earth, with small herbs like *Carex digitata* L., *Melampyrum silvaticum* L., and *Fragaria vesca* L.

7. *Tall herbs:* spruce forest on brown earth, with tall herbs like *Filipendula ulmaria* (L.) Maxim., *Athyrium filix-femina* (L.) Roth., and *Aconitum septentrionale* Koelle.

2.1.3 Study area for mites in decomposing sporocarps

Dead sporocarps of different wood-living polypore fungi were sampled in an old spruce forest in the Østmarka area, about 20 km east of Oslo [24, 25].

2.1.4 Study areas for mite activity under and within snow

Activity under snow was studied in an old spruce forest with *Vaccinium myrtillus* vegetation near Veggli in Numedal valley, about 150 km NW of Oslo. Here, at 850 m above the sea level, a snow cover of 1–2 m is common [23]. In the main study area, Nordmoen, mite activity was studied both under and within snow [11].

2.2 Methods for field studies

2.2.1 Soil sampling

Each vegetation type in areas A and B was sampled twice, in autumn 1977 and in spring 1978. Using a soil corer of 10 cm^2, 20 soil cores were taken both

Figure 1.
The poorest coniferous forest type: slow-growing pines on a thin soil layer dominated by Cladonia lichens. Certain drought-tolerant, lichen-feeding mites were abundant here. Photo: S. Hågvar.

during spring and autumn in each vegetation type. The cores were divided into 0–3 and 3–6 cm depth. In the main study area at Nordmoen, the same sampling method was used. Here, a clearcut area with 0.5 m high *Picea abies* seedlings was chosen for intense studies. Eight random replicates were established, each 4 × 4 m. Density of mites per replicate was based on 10 soil cores, each 5.3 cm^2 and 6 cm deep.

2.2.2 Artificial acidification and liming

Lime was applied as crushed $CaCO_3$ (3000 kg Ca0 ha^{-1}), and 50 mm of artificial acid rain was applied monthly by adding sulfuric acid to ground water (**Figure 2**). Treatments were no watering, pH 6 (control), pH 4, pH 3, pH 2.5, and pH 2. The natural pH in the organic layer (upper 3 cm) was 3.9. Liming increased pH about 2 units, and the strongest acid reduced pH about 0.5 units. Only application of acid rain with pH 3 or stronger lowered the pH in the organic layer [14].

2.2.3 Litter bag studies on succession

The clearcut area in the main study area was used to study the mite succession during decomposition of spruce needles [19] and birch leaves [12, 13]. Cylindrical litter bags, 3 cm high and with a diameter of 3.4 cm, were filled with 4.2 g (dry weight) of naturally shed spruce needles. The litter bags were then inserted into holes made in the raw humus layer, which had a corresponding depth. This is not a natural position of the litter, but it allowed to study the preference among mites for different decomposition phases. While the litter bags stood in this fixed position, in contact with various depth levels of the organic horizon, all species had a continuous access to the needles. With a mesh size of 0.6 mm, migration to and from the bags was easy for all microarthropods.

Succession in decomposing birch leaves was studied in a similar way in the same site. Cylindrical litter bags with a mesh size of 1 mm, 3 cm high and with a diameter of 6.5 cm, were each filled with 6.85 g (dry weight) of naturally shed birch leaves. These bags also received artificial rain of pH 6, 4, 3, and 2.

Figure 2.
Artificial acid rain is applied on a 4 × 4 m experimental plot with small pine trees.

2.2.4 Mites living inside decomposing spruce needles

In the main study area at Nordmoen, naturally shed spruce needles were sampled on snow and dried. Later, needles were stuck into fine-meshed nylon strips, which were placed on the ground of a 10–20 m high spruce stand. Gradually, needles were covered by new litter in a natural way. Strips with needles were recovered after 4, 12, 16, 24, 35, 38, 40, and 52 months [22].

2.2.5 Sporocarp sampling and extraction

Dead sporocarps were brought to the laboratory, carefully fragmented, and mites were extracted in funnels, using heat from a light bulb [24, 25].

2.2.6 Sampling mites active under snow

Specially designed pitfall traps were used [23]. The mechanism allowed sampling without disturbing the subnivean air space near the traps.

2.3 Methods for laboratory studies

2.3.1 A "preference" experiment

This was a greenhouse experiment, where forest soil was kept in large plastic boxes [10]. Microarthropods (and microflora as well) had the opportunity to colonize sterilized soil (raw humus, poor mull, and rich mull) which had been adjusted to three different pH levels. Cylindrical litter bags with a mesh size of 1 mm, 3 cm high and with a diameter of 6.5 cm, were used. The design can be characterized as a preference experiment, where also the ability to reproduce during the four-month period influenced the establishment of each species.

2.3.2 Microcosm studies

Small microcosms were used, consisting of a cylindrical, open litter bag which was inserted into a lidded plastic container. The litter bag was 3 cm high, 3.4 cm in diameter, and made from a nylon cloth with 0.6 mm mesh size. Holes drilled in the plastic container were covered with nylon cloth of 5-μm mesh size. Before adding microarthropods to sterilized soil, microflora was introduced partly by soil water sieved through 5-μm pores and partly by allowing soil fungi to grow in through corresponding pores for 1–2 months. Then animals were added, either from monocultures or from ordinary soil samples [17]. Raw humus adjusted to different pH levels was used in the microcosms. About 25 microcosms were extracted after 3, 6, and 12 months, respectively. This setup allowed for studying the effect of soil pH on population growth in monocultures of selected species. An interesting by-product was the effect of predatory Gamasina mites, which survived in some microcosms, but went extinct in others [21].

3. Results and discussion

3.1 The coniferous forest floor: a high density and species rich habitat for mites

Podzol soil with vegetation type 4 in the main study area contained 48 species of Oribatida and 12 species of Mesostigmata (**Table 1**). The density of mites was high.

In the upper 6-cm soil, the mean numbers per m^2, based on eight replicates, were: Prostigmata (Actinedida) 490,000, Oribatida 220,000, and Astigmata (Acaridida) 10,000. The total mite density was 720,000 per m^2. The highest total density in one replicate amounted to 1.2 million mites per m^2 [14].

Comparable data exist from Finland and Sweden. In southern and central parts of Finland, mites were studied in four coniferous forest sites [28]. The localities corresponded to vegetation types 2 and 4 in the present study. The densities of oribatids, 186,000–351,000 per m^2, were in the same order of magnitude as in

Oribatida	**Oribatida (continued)**
Adoristes poppei (Oudemans)	*Oppia subpectinata* Willmann
Autogneta parva Forsslund	*Oppia unicarinata* (Paoli)
Autogneta trägårdhi Forsslund	*Oppia nova* (Oudemans)
Belba cf. *compta* Kulczynski	*Oribatula tibialis* (Nicolet)
Brachychochthonius zelawaiensis (Sellnick)	*Palaeacarus* sp.
Caleremaeus monolipes (Michael)	*Parachipteria* cf. *willmanni* (V. D. Hammen)
Camisia biurus (C. L. Koch)	*Paraleius* cf. *leontonycha* (Berlese)
Camisia cf. *lapponica* Trägårdh	*Paulonothrus longisetosus* (Willmann)
Camisia spinifer (C. L. Koch)	*Pergalumna nervosus* (Berlese)
Carabodes femoralis (Nicolet)	*Phthiracarus* sp.
Carabodes forsslundi Sellnick	*Platynothrus peltifer* (C. L. Koch)
Carabodes labyrinthicus (Michael)	*Porobelba spinosa* (Sellnick)
Carabodes marginatus (Michael)	*Scheloribates laevigatus* (C. L. Koch)
Carabodes subarcticus Trägårdh	*Steganacarus* sp.
Cepheus cepheiformis (Nicolet)	*Suctobelba subcornigera* (Forsslund)
Ceratozetes sp.	*Tectocepheus velatus* (Michael)
Chamobates incisus (V. D. Hammen)	*Zygoribatula* cf. *trigonella* Bulanova & Zachvatkina
Chamobatidae sp.	**Mesostigmata**
Eueremaeus silvestris (Forsslund)	*Eviphis ostrinus* (Koch)
Eupelops duplex (Berlese)	*Gamasellus montanus* (Willmann)
Eupelops geminus (Berlese)	*Hypoaspis forcipata* Willmann
Euphthiracaridae	*Leioseius bicolor* Berlese
Hemileius initialis Berlese	*Parazerkon sarekensis* Willmann
Hypochthonius rufulus C. L. Koch	*Pergamasus* cf. *lapponicus* Trägårdh
Liacarus cf. *coracinus* (C. L. Koch)	*Pergamasus parrunciger* Bhattacharyya
Licneremaeus licnophorus (Michael)	*Pergamasus robustus* Oudemans
Nanhermannia cf. *forsslundi* Karppinen	*Prozercon kochi* Sellnick
Nothrus silvestris Nicolet	*Trachytes* sp.
Oppia cf. *translamellata* (Willmann)	*Veigaia cerva* (Kramer)
Oppia obsoleta (Paoli)	*Veigaia nemorensis* C. L. Koch
Oppia ornata (Oudemans)	

Table 1.
In the clearcut area of the main study site Nordmoen, 48 species/taxa of Oribatida were recorded, and 12 of Mesostigmata.

the present study for vegetation type 4. However, their Prostigmata densities, 34,000–80,000 per m^2, were only about one tenth of ours. As much as 62 oribatid taxa were recorded in a Finnish spruce site with vegetation type 4. In another Finnish study of spruce forest soil, 35 taxa of oribatids were recorded and a relatively low density, only 70,000 oribatids per m^2 [29].

In an old Swedish pine forest of vegetation type 1–2, 52 oribatid species were recorded and very high densities [30]. As much as 425,000 oribatids per m^2 were found, which surpasses both the Norwegian and Finnish densities mentioned above. Prostigmata numbers (210,000 per m^2) were between Norwegian and Finnish numbers, and total mite numbers (684,000 per m^2) approached the high Norwegian number of 720,000. We can conclude that Nordic coniferous forest soils with raw humus have a very rich mite fauna, both in oribatid species and in total mite numbers.

3.2 Horizontal and vertical distribution

The main study area had very homogeneous soil conditions over a large area. It was a flat plain with stone-free, sandy soil, without visible variations in moisture conditions or vegetation. Still, as shown in **Table 2**, the horizontal distribution of many species showed considerable local variations [14, 15].

In another experiment, litter bags with birch leaves were placed in the humus layer of four random blocks. The mite fauna which colonized the litter varied significantly between blocks [12]. The Astigmata species *Tyrophagus* cf. *fungivorus* (Oudemans) colonized heavily in Blocks 1 and 2, while *Oppia ornata* occurred mainly in the other two. Actinedida mites were especially numerous in litter bags of Block 4, while the same litter bags had the lowest number of *Autogneta trägårdhi*. Block 1 had high numbers of *Oribatula tibialis*, while *Chamobates incisus* had its highest numbers in Blocks 2 and 3 (**Table 3**).

The study of vertical distribution in mites was restricted to the upper 6 cm. *Carabodes* species only rarely occurred in the 3–6 cm layer and were to a large degree

Species	Group	Densities
Parazercon sarekensis	M	1.7–5.2
Veigaia nemorensis	M	0.1–1.7
Tectocepheus velatus	O	20–110
Nothrus silvestris	O	2–95
Brachychochthonius zelawaiensis	O	2–100
Oppia obsoleta	O	0–5.5
Oppia nova	O	0–4.5
Paulonothrus longisetosus	O	0–3.7
Brachychthoniidae	O	20–200
Total Oribatida		80–360
Astigmata (Acaridida)		3–30
Prostigmata (Actinedida)		230–850
Total Acari		400–1200

Table 2.
Lowest and highest density of various mites (1000 per m^2) in eight random study plots (each 4 × 4 m) on a flat and homogeneous forest area. O = Oribatida and M = Mesostigmata. Mite density in a given plot was the mean of 10 soil cores, 6 cm deep and with a surface area of 5.3 cm^2.

Species	Sample No	Block numbers				Significance
		B 1	B 2	B 3	B 4	
Tyrophagus cf. *fungivorus*	I	533.5	735.8	13.5	2.3	B3 & B4 < B1 & B2
Oppia ornata	III	0	0	21.7	6.3	B3 > B1, B2 & B4
Prostigmata (Actinedida)	II	46.8	77.5	211.0	343.8	B4 > B1 & B2
Autogneta trägårdhi	I	56.6	45.1	55.8	21.4	
Oribatula tibialis	II	208.5	92.7	44.9	28.6	B1 > B2, B3 & B4
Chamobates incisus	II	0	2.0	4.6	0.6	B3 > B1 & B4

Table 3.
Examples of how the number of mites per litter bag with birch leaves may vary between four blocks in a flat and apparently homogeneous forest floor [12].

living in close connection with *Cladonia* lichens on the surface [25]. In the main study area, there was no sharp change in the mite fauna between the organic layer (0–3 cm) and the bleached mineral layer (3–6 cm). For instance, the large *Nothrus silvestris* was equally abundant in the two layers. However, the addition of strong doses of lime or artificial acid rain was apparently stressful for several mites, forcing animals to deeper layers. After treatment, the following oribatids moved significantly deeper, shifting from living mainly in the organic layer, to live mainly in the mineral layer: *Nothrus silvestris, Suctobelba* sp., *Brachychochthonius zelawaiensis*, and total oribatids. However, Prostigmata mites showed a shift upwards in the soil profile [14]. A frequent natural stress factor in soil is drought. In a Finnish forest, *Nothrus silvestris* was seen to migrate into deeper layers during warm periods [31].

In the comparative study between different vegetation types and soils, all the six selected mites showed variations in depth distribution, not only between habitats, but also between seasons [15]. On the average, the following percentages of the populations occurred in the upper 3 cm compared to 3–6 cm depth: 85% in *Tectocepheus velatus*, 65% in *Parazercon sarekensis*, 60% in *Schwiebea* cf. *cavernicola* Vitzthum, 54% in *Brachychochthonius zelawaiensis*, 52% in *Nothrus silvestris*, and 51% in *Schwiebea* cf. *nova* (Oudemans). The somewhat deeper distribution of *Nothrus silvestris* compared to *Tectocepheus velatus* has been confirmed by other studies [32–34].

3.3 Effect of ground vegetation and soil type

Eight mite species were studied systematically with respect to vegetation types and soil characteristics [15, 25]. Five belonged to the oribatids, two belonged to Acaridida, and one to Mesostigmata (**Table 4**). Most species preferred poor and acidic podzol soils with raw humus (up to vegetation type 4), but *S.* cf. *cavernicola* had the highest density in a poor brown earth (type 6). None of the eight species were abundant in the richest soil, a brown earth with mull humus (type 7). The non-*Carabodes* species in **Table 4** were tested for correlation between population size and soil chemical parameters. Soil pH, and the accompanying parameters base saturation and calcium content, turned out to be the strongest explanatory factor.

Some other *Carabodes* species were so rare in all soils that they have been excluded from **Table 4**, but further mentioned under the next point.

Comparable data from Finland and Sweden confirm that *Nothrus silvestris* and *Tectocepheus velatus* occur in many different plant communities of coniferous forest, but typically in acid raw humus, and with low densities in richer soils [31, 33, 35, 36]. Although preferences exist, it has been concluded on a general basis that many oribatid species are able to persist in a wide range of humus forms and vegetation types [37].

Species	Group	Vegetation type						
		1	2	3	4	5	6	7
Carabodes subarcticus	Oribatida	15.2	0.5	2.5			0.02	
Carabodes willmanni	Oribatida	37.1	43.1	4.7				
Parazercon sarekensis	Mesostigmata	3.4	1.5	1.8	4.0	2.9	1.3	
Tectocepheus velatus	Oribatida	175.3	66.8	99.1	47.2	7.4	11.5	0.7
Brachychochthonius zelawaiensis	Oribatida	0.9	1.5	27.7	38.8	12.2	1.6	0.5
Nothrus silvestris	Oribatida	1.8	3.5	14.8	22.0	3.7	7.3	1.9
Schwiebea cf. *cavernicola*	Acaridida	0.7	1.3	7.1	4.0	7.8	11.3	2.6
Schwiebea cf. *nova*	Acaridida	0.4	0.9	1.8	12.2	3.4	0.4	0.5

Numbers are mean value from two localities, each sampled during spring and autumn. Vegetation types 1–7 are described in Material and Methods. Soil fertility increased from left to right. For complete vegetation data, see [27].

Table 4.
Abundance (1000 per m² in the upper 6 cm soil layer) of some common mite species in seven different vegetation types in coniferous forest.

3.4 Carabodes: a genus with different life forms

The combined study of mites in different coniferous forest types and mites in decomposing polypore fungi illustrated that closely related species within a genus (*Carabodes*) can fill quite different niches in the forest ecosystem [25]. The most common *Carabodes* species in soil were rare in sporocarps and vice versa. The first two species in **Table 5** were considered *Cladonia*-feeders on the ground and were able to live in a dry forest floor. The third species on the list is also a lichen-feeder, which often climbs tree stems. Then, we have three fungal feeders which decompose dead sporocarps and may achieve high densities in these patchy and temporary habitats. Their relative numbers were rather similar in dead sporocarps of five different fungal species, including annual and perennial sporocarps, soft and hard. Although being tolerant to different fungal species, these specialists were considered vulnerable in forests with little dead wood and few sporocarps [25]. The five lower species have been found in low numbers, both in sporocarps, in dead wood, and in soil. They are either generalists or have unknown preferences.

3.5 Effect of soil acidity on mites—natural and manipulated

3.5.1 General results

Three approaches were used to test whether soil pH was an important environmental factor for mites. First, a "preference experiment" was arranged in the laboratory [10]. Here, mites were allowed to colonize soils adjusted to different pH levels. Second, we studied responses to artificial pH changes in soil through liming and artificial acid rain, both in the field and in the laboratory [13, 14]. Third, mites were sampled in natural soils of varying pH, to check if there were species that occurred mainly at certain pH levels [15].

Table 6 gives the most consistent results from the first two approaches. Clear responses were found in three oribatid species, in total Oribatida, and in the Acaridida species *Schwiebea* cf. *nova.* Raised pH due to liming reduced densities of these taxa, while acidification usually led to higher densities. The third approach from natural soils of different pH supported the pattern: species which increased

Species	In sporocarps	In dead wood	In soil	Remark
C. willmanni Bernini	(+)		++++	*Cladonia*-feeder on the ground
C. subarcticus Trägårdh	(+)	+	++	*Cladonia*-feeder on the ground?
C. labyrinthicus (Michael)	+(+)	+	+(+)	Lichen-feeder, common on tree stems
C. femoralis (Nicolet)	++++	++	+	Polypore specialist
C. areolatus Berlese	+++	++	(+)	Polypore specialist
C. reticulatus Berlese	+++	+		Polypore specialist
C. marginatus (Michael)	(+)	+	+	
C. forsslundi Sellnick	+	+	+	
C. rugosior Berlese	+	+	(+)	
C. tenuis Forsslund	+	+	(+)	
C. coriaceus Koch	+	+	(+)	

Very high abundance is subjectively indicated by ++++ and very low abundance by (+). Short remarks are given for some species.

Table 5.
Simplified overview on the occurrence of various Carabodes species in different forest habitats, compiled from several sources. From [25].

Species	Effect of liming		Effect of acidification			
	Colonization experiment [10]	Field experiment [14]	Colonization experiment [10]	Field experiment [14]	Birch leaves [13]	
					Field	Green-house
Nothrus silvestris	—	—	+			
Tectocepheus velatus	—	—	+	+	+	+
Brachychochthonius zelawaiensis	—	—	+		+	—
Total Oribatida		—	+	+	+	
Schwiebea cf. *nova*	—		+			+

Table 6.
Significant effects of liming and acidification on mite densities. Compiled from several studies.

in numbers during artificial acidification were often numerous in naturally acid soils [15]. It was concluded that soil pH was a highly relevant environmental factor for certain mites. Among them was the rather large oribatid species *Nothrus silvestris* (**Figure 3**).

In field experiments with application of artificial rain, the structure of the mite community changed in a characteristic way. **Figure 4** shows how the dominance structure was influenced by liming and application of "rain" with pH 2.5 and 2. Watering with pH 6 was considered as control. The dominance of Oribatida increased with increased acidification. Changes were mainly due to reactions in the sensitive species from **Table 6**.

Finnish [38] and Swedish [39] experiments conformed well with these data, as well as other studies referred to in [14].

Soil acidity is, of course, only one of many factors that modify the abundance of these species, and the relation is not absolute. Even if the pH level is favorable, other limiting factors, for instance drought, may depress populations. The

Figure 3.
Nothrus silvestris is an oribatid species that is typical for acid raw humus and declined after liming. Photo by courtesy of SNSB—Zoologische Staatssammlung München.

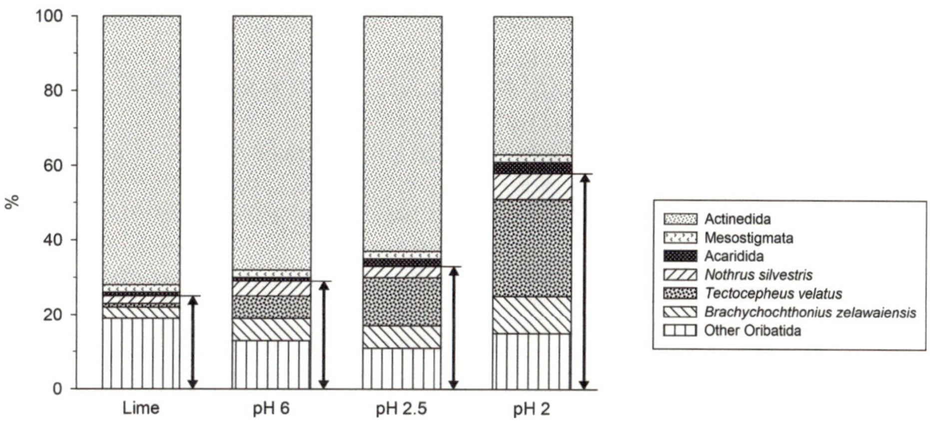

Figure 4.
Effect of liming and acidification (the two most extreme treatments, pH 2.5 and 2) on the relative dominance among mites. Watering with pH 6 was considered as "control." The dominance of Oribatida mites, indicated by double arrows, increased with increased acidification [14]. Actinedida = Prostigmata and Acaridida = Astigmata.

experiments support the following conclusion: a high abundance in certain species can only be achieved within a certain pH interval (and only if other factors are not limiting), while within another pH interval, high abundance cannot be achieved. In soils of the latter pH interval, the acidity level (or correlated factors) seems to be limiting [16].

3.5.2 Is competition a key factor?

Relations between abundance of mites and soil acidity are difficult to explain. Soil pH is a measure of the H^+ activity of the soil solution. This parameter may have a direct importance for the water-living part of the soil fauna (such as Protozoa and Rotifera), and to other groups living in contact with the soil solution, as Nematoda [40, 41]. Both Enchytraeidae and Lumbricidae prefer relatively high moisture in the soil [42]. The survival of the Enchytraeidae species *Cognettia sphagnetorum* Vejdovsky decreased rapidly when the animals were submerged in diluted sulfuric acid of pH below 4 [9]. Many Enchytraeidae species show distinct relations to soil pH, both in experiments and in the field [10, 39, 43, 44]. The dependence of Lumbricidae species upon soil pH is well documented [45–47].

Microarthropods, on the other hand, have a hydrophobic cuticula and are restricted to the air-filled pore spaces of the soil. The relations described are probably indirect. Several possibilities have been discussed [16]: changes in ground vegetation due to artificial acid rain, direct effects of lime or sulfuric acid, various factors correlated to soil pH, changed predation pressure, availability of fungal hyphae as food, or fecundity. After having refuted several hypotheses, the following laboratory experiment pointed toward competition as a possible explanation [18].

Some microcosms were added a full soil fauna, while others were monocultures of selected species. The acidophilic Acaridid mite *Schwiebea* cf. *nova* (later named *S.* cf. *lebruni* Fain) thrived in monocultures. Starting with 30 specimens, populations increased to around 2000. Surprisingly, population growth in monoculture was lowest in the most acid soil. In the "full fauna" microcosms, however, the species revealed its typical acidophilic character and achieved the highest populations in the most acid soil. Quite parallel results were achieved for the acidophilic springtail *Mesaphorura yosii* (Rusek) [18]. These species have an optimum at a high pH when being alone. However, by some reason, they seem to be good competitors at low pH. They were winners both in natural soils with a low pH and in various experiments with artificial acid rain. Also the acidophilic oribatid *Nothrus silvestris* reproduced best in limed soil when alone [18].

For Collembola, other laboratory studies on population growth, with or without other species present, have illustrated that competition occurs [48, 49]. In most cases, the presence of another species reduced population growth. The most common mechanism was disturbance during oviposition. A classic study about competition among oribatid mites was performed in microcosms with natural soil. Two species with overlapping niches, *Hermaniella granulata* (Nicolet) and *Nothrus silvestris*, were first bred in monocultures. When put together, both species underwent significant shifts in their use of space and food. Their vertical distribution changed so that *Hermaniella* moved upwards into the litter layers, while the *Nothrus* population increased in the deeper fermentation layer [50].

Competition may attain many forms, and the topic is not easy to disentangle. However, since species live so densely packed in soil, one can imagine that disturbance or limited space or food may have an influence. If competition is a key factor regulating population size in soil, a general study of competition in microarthropods might be rewarding. Although a species may have its set of preferences, the key quality may be its ability to compete under suboptimal conditions.

3.6 The effect of predatory Mesostigmata mites

While the function of soil mites is often focused on their role in decomposition, predatory Mesostigmata mites have the potential to control the density of little sclerotized prey of various taxa. The evolution of strongly sclerotized bodies in many oribatid species obviously has an antipredator role.

The microcosm experiment described above illustrated the predatory effect of large Gamasina mites. At the start, 96% of the cultures contained predatory Gamasina mites, mainly *Veigaia nemorensis*. This percentage was reduced to 73% after 3 months, 62% after 6 months, and 50% after 12 months. The local extinction of these predators often resulted in very high densities of springtails or mites. For instance, after 1 year, the number of *Schwiebea* mites in certain predator-free microcosms could amount to several hundred, while predator-containing cultures usually had numbers below 30. Also for springtails, the highest populations were recorded in cultures where predatory Gamasina mites had gone extinct [21].

From the literature, another laboratory experiment illustrated well this top-down control of microarthropods. The addition of predatory mites to isolated soil cores containing a natural microarthropod fauna reduced the density of small and less sclerotized oribatids, as well as Collembola and Protura [5].

In agroecosystems, edaphic Mesostigmata have been shown to be important predators of Collembola and Nematoda, and those living on plants may efficiently control pests like spider mites [4].

3.7 Succession in the mite community during decomposition of spruce needles and birch leaves

In the main study area at Nordmoen, the clearcut area was used for litter bag studies, as described above. Litter bags with birch leaves were placed out in July 1975. There were four samplings: September 1975, April 1976, September 1976, and November 1978. The number of leaf-containing litter bags harvested at each sampling was 32, 68, 128, and 78, respectively.

Litter bags with spruce needles were placed out in September 1977, and samplings were made after 7 weeks, 8 months, 1 year, 2 years, 5 years, and 10 years. All samplings, except for the second one, were taken at the same time of the year. There were four replication sites, and 5–15 litter bags were harvested from each replication at a given sampling. Detailed results were given for birch leaves [12] and for needles [19]. Here, the main trends shall be presented and compared.

In both litter types, a gradual change in the mite community was observed during the decomposition process. However, the succession pattern differed in spruce needles and birch leaves. It means that mites in the surrounding soil were selective about which litter they colonized, at which rate, and at which decomposition stage. For instance, two oribatid species which were common in the soil, *Tectocepheus velatus* and *Nothrus silvestris*, never became abundant in litter bags. On the other hand, certain low-density species in soil could achieve very high densities in the bags. In such cases, a high density was only seen in one of the litter types. Examples in spruce needle bags were high density of *Eremaeus* sp. after 1 year, *Steganacarus* sp. after 5 years, and *Oppiella nova* after 10 years.

A considerable number of spruce needles were decomposed from the inside by certain specialized oribatid mites [22, 26]. Smaller, deeper-living species became abundant after 5–10 years, when the needles had been more or less fragmented. The fragmentation created new microhabitats and perhaps allowed for a more intense microfloral colonization.

While colonization of needle litter was slow, and no species or group achieved its maximum abundance within 8 months, colonization of birch leaves was much faster. Here, certain mites, which had a low density in the surrounding soil, appeared very numerous already after 7 weeks. Examples were three oribatid mites: *Oribatula tibialis*, *Eupelops duplex*, and *Autogneta trägårdhi*, and one Acaridida (Astigmata): *Tyrophagus* cf. *fungivorus*. Studies of the gut contents of these four species revealed a mixture of fungal spores and hyphae, and some guts contained mainly spores. This indicated an intense grazing, probably due to a temporal "flush" of fungal activity. The same was seen for certain springtail species [12]. It is, of course, important for soil microarthropods to detect such spatial and temporal food sources, and it is reasonable to assume that animals were attracted from surroundings by smell. Also other studies have documented a rapid migration of microarthropods into decomposing deciduous leaves [51–53]. Such species can be characterized as mobile opportunists. An abundant food source may allow a high number of species and specimens to coexist in a substrate with a low structural diversity. The body of *Eupelops duplex*, but also other species, was often covered by fungal spores or hyphae, promoting the spread of microflora to all parts of the litter. The study also indicated that several species did not reproduce in the substrate, but only visited it during the adult stage for feeding purpose.

Table 2 shows that litter-dwelling pioneer mites in birch litter had a very uneven horizontal distribution, within 20–50 m. It meant that the succession pattern in the early decomposition phase varied widely, even within an apparently homogeneous forest floor. In later decomposition stages, however, the microarthropod community was less variable and more predictable.

In both litter types, large, surface-living species were among the early colonizers, while smaller, usually deeper-living species, took over the dominance in later decomposition stages. Since the litter bags had continuously contact with the whole organic layer in the actual soil, the succession studies confirmed that deeper-living, and often small species, preferred a more decomposed material.

While this experiment demonstrated that species often had different preferences for litter type or decomposition stage, it also showed that many species had wide tolerances and could survive, sometimes in low densities, under rather different circumstances. In an English study of oribatid mites in decomposing leaves of beech and chestnut, the 12 most abundant species were present in the litter bags throughout the 20-month study period. During this time, species were able to remain by changing their feeding habits [51]. Another example of high tolerance among oribatid species to different decomposition stages of leaf litter is from Central Amazonas. During the one-year long study, there was no successional changes in the species composition [54].

Few decomposition studies last long enough to describe the late stages of the microarthropod succession. For instance, in a study of root litter decomposition, it was found that oribatid mites showed a preference for the late stages of decomposition [55]. A general challenge in litter bag studies is how to simulate natural conditions. And even if natural conditions are achieved, the result may only have local value. Anyhow, due to a high species number and an ecological flexibility in many species, mites do in several ways contribute in transforming litter to humus. This is exemplified in the next chapter.

3.8 From litter to humus: can mites influence the process and the products?

Juveniles of certain specialized mites excavated cavities in about 40% of newly fallen spruce needles. Their activity reduced the decomposition rate of the actual needles, at least temporarily, probably because their excrements decomposed slowly [22, 26]. The adult mites, which hatched after about 2 years, attacked other needles from the outside and fragmented these (**Figure 5**). Their “inert” excrement pellets

Figure 5.
Two spruce needles that have been fragmented by adult “box mites” (Steganacarus cf. striculus) kept in culture. Two ellipsoide-shaped animals are seen. Excrement pellets are numerous. Photo: S. Hågvar.

may contribute to a stable humus layer and perhaps to carbon sequestration [26]. Also other studies have pointed to the fact that fecal pellets of oribatids decompose slowly and may contribute significantly to humus production [56, 57]. Even pine needles can be tunneled by phthiracarid mites [57].

Individual spruce needles may show quite different decomposition patterns, even if situated close to each other in soil. While some are heavily transformed to excrement pellets, others remain morphologically intact for years. Needles which happen to come in close contact with fine roots may be rapidly "dissolved." These "individual fates" of needles may explain the heterogeneous structure of deep humus [26].

3.9 Mite activity beneath and within snow

Norwegian coniferous forest is covered by snow for several months each year. When the snow layer exceeds about 20 cm, the temperature at the soil surface stabilizes around 0°C [58]. At this temperature, several surface-living invertebrates are active in the subnivean air space and even feeding [23]. Among these are several species of springtails and mites. During two winters, pitfall traps were operated under 30–150 cm snow in a high altitude spruce forest with bilberry vegetation in Southern Norway. Traps were emptied and replaced at least monthly during the snow-covered period from October/November to April/May.

Twelve taxa of Oribatida were trapped and 10 of Mesostigmata. A number of Prostigmata were also taken (**Table 7**). The Oribatida material was dominated by one species, *Platynothrus capillatus*. All developmental stages of this species were active under snow, and fungal hyphae and spores in their guts proved winter feeding. It was assumed that they were grazing on certain fungi known to decompose litter beneath snow (snow molds) [23]. Also other species of Oribatida, as well as some Prostigmata, had visible gut content.

In the main study area at Nordmoen, microarthropod activity both beneath and within snow was studied [11]. Most surface-living springtails were winter active and even migrated up into the snow layers. Among mites, four predacious Mesostigmata mites and one oribatid species (*Adoristes poppei* Oudemans) were taken in small numbers in pitfall traps, together with numerous Prostigmata. Mites were also found within the snow layers: some Prostigmata, seven taxa of predacious Mesostigmata, and six taxa of oribatids, of which *Adoristes poppei* was the most numerous. It was suggested that microarthropods went into snow to escape possible harmful water logging or ice formation in late winter [11].

3.10 Remarks on ecological flexibility and vulnerability

Several mite species showed a high tolerance for different plant communities, soils, humus types, litter type, and succession phase. Both birch leaves and spruce needles in litter bags were colonized by a high number of oribatid species. Several of them occurred in both substrates, although colonization was much slower in needle litter. Birch leaves represented an uncommon substrate at the actual site, but probably offered a flush of fungal food. Furthermore, at least some individuals of most species participated in various decomposition phases, where the substrate underwent significant changes. Except for pH, mites seemed to have few strong relations to soil chemical parameters [15].

Each mite species continually adjusts its vertical position, as far as narrow pores allow, to optimize its survival, food access, and reproductive ability. Such changes were seen also in the horizontal distribution. A more fixed vertical or horizontal position of each species could reduce interspecific competition but would be a disadvantage as soon as adverse or favorable conditions developed in certain layers or sites.

ORIBATIDA	**Stage**	**Number trapped**	**Gut contents observed?**
Camisia biurus	Ad	3	No
	T	2	Yes
	P	3	No
	L	2	No
Carabodes labyrinthicus	Ad	2	Yes
Carabodes marginatus	Ad	1	No
Carabodes sp.	T	1	No
Chamobates pusillus (Berlese)	Ad	1	No
Eobrachychthonius borealis Forsslund	Ad	3	Yes
Oppiella neerlandica (Oudemans)	Ad	12	Yes
Oppiella sp.	Ad	1	Yes
Oribatella calcarata (C.L. Koch)	Ad	1	No
	D	7	Yes
	P	5	Yes
Platynothrus capillatus (Berlese)	Ad	10	Yes
	T	9	Yes
	D	14	Yes
	P	8	Yes
	L	2	Yes
Steganacarus sp.	Ad	1	Yes
Belba sp.?	Ad	2	No
	T/juv	5	Yes
MESOSTIGMATA			
Mixozercon serlachii Lehtinen		1	
Zercon curiosus Trägårdh		1	
Zercon colligans Berlese		2	
Holoparasitus sp.		1	
Lysigamasus lapponicus (Trägårdh)		4	
Vulgarogamasus kraepelini (Berlese)		14	
Veigaia nemorensis		6	
Trachytes aegrota (C.L. Koch)		2	
Urodiaspis tecta (Kramer)		1	
Uropodina sp., nymph		2	
PROSTIGMATA (ACTINEDIDA)		115	Yes
TOTAL		244	

Numbers per 12 functioning traps. Only periods with a continuous snow cover are included. Ad = adults, T = tritonymphs, D = deuteronymphs, P = protonymphs, and L= larvae.

Table 7.
Mites (Acari) caught in pitfall traps under snow during two winter seasons in a high altitude spruce forest, central South Norway. Modified from [23].

The present documentation [12] showing that many springtails and mites change their food habits through the different successional stages is in good accordance with other observations [51].

Within both springtail and mite communities, it is a general pattern that most species are relatively rare. A high tolerance for various habitat or nutritional factors, often combined with asexual reproduction, may keep species going on in low numbers. However, when special conditions are created locally, rare species may act as opportunists and flourish temporarily. They also represent an important resource if the ecosystem has to adapt to a new situation, for instance due to climate change.

Although single species may show tolerance to different environmental conditions, the mite community as a whole can be vulnerable to various types of human disturbance. For instance, in New York, the diversity of oribatid mites decreased along a gradient of land use types in the order from forests, via abandoned fields and willow, to corn [7]. A European review on mites as indicators of soil biodiversity and land use monitoring illustrated how sensitive mite communities can be to various types of soil disturbance [59]. Changes in the dominance structure of mite communities were suggested to be an "early warning criterion" for stressed mite communities. The author concluded that residual natural and semi-natural habitats (such as old woodlands, riparian ecosystems, old hedges, and grasslands) with species-rich mite communities found in rural and urban landscapes should be preserved as refuges for dispersion of soil fauna.

4. Conclusions: spotlights in short

Coniferous forests are rich in mites: a podzol soil with acid raw humus may contain more than a million mites per m^2. This includes a species-rich oribatid fauna.

Flexible vertical and horizontal distribution: mites can adjust both their depth in the soil profile and their horizontal distribution, either to escape stress or to aggregate in a patchy and temporary food source.

Opportunism as a successful strategy: several litter-dwelling mite species rapidly colonized birch leaves in an early decomposition phase, in order to feed on a temporary and patchy flush of fungal hyphae and spores.

Substrate flexibility: decomposition of spruce needles and birch leaves followed quite different succession patterns, but several mite species participated in both. Closely related species may differ widely in habitat choice and life forms: this was exemplified in the genus *Carabodes*.

Predacious Gamasina mites matter: microcosm studies showed high population growth of certain mites and springtails if predatory Gamasina mites went extinct.

Oribatids matter in the decomposition process from litter to humus: specialized oribatids excavate spruce needles and produce slowly decomposable excrements.

Soil acidity matters: colonization experiments and population studies in monocultures showed that soil pH affected population size in certain species. This led to predictable changes in the community structure of mites.

Successful competition under suboptimal conditions: surprisingly, certain mites common in acid soils thrived best in less acid soil when being alone (in monoculture). However, in acid soil, they were good competitors.

Mites are winter active: several mites are active under snow, often feeding. Some even penetrate into the snow layer.

5. Final remarks

There is an increasing awareness for preserving the huge biodiversity of soils [1, 60, 61]. Fragmentation and various management practices of forests may affect even these tiny animals. Some microarthropod species are confined to local soil types, for instance under dry or wet conditions. Furthermore, a forest contains various microhabitats in addition to soils. Examples are moss or lichen vegetation on certain trees, suspended soils in birds´ nests, mold in old, hollow trees, decomposing wood, or fruiting bodies of various fungi. To preserve the species, diversity of microarthropods may demand a relatively large forest area, covering a variety of vegetation types, soils, humus types, and microhabitats.

Due to their long life span, low fecundity, slow development, and low dispersion ability, oribatid mites have been suggested as suitable indicators of soil biodiversity and land use monitoring. In this respect, there is a need to develop standardized procedures for sampling and data analysis [59].

Acknowledgements

I am grateful for being allowed to reuse **Figure 4** from Oikos, **Table 5** from Scandinavian Journal of Forest Research, and **Table 7** from Soil Organisms. Zoologische Staatssammlung München gave permission to use the photo of *Nothrus silvestris.* Ole Wiggo Røstad kindly helped with some figures.

Author details

Sigmund Hågvar
Faculty of Environmental Sciences and Natural Resource Management, Norwegian University of Life Sciences, Ås, Norway

*Address all correspondence to: sigmund.hagvar@nmbu.no

References

[1] Wall DH, Bardgett RD, Kelly EF. Biodiversity in the dark. Nature Geoscience. 2010;**3**:297-298

[2] Siepel H, Maaskamp F. Mites of different feeding guilds affect decomposition of organic matter. Soil Biology and Biochemistry. 1994;**26**:1389-1394

[3] Schneider K, Migge S, Norton RA, Scheu S, Langel R, Reineking A, et al. Trophic niche differentiation in soil microarthropods (Oribatida, Acari): Evidence from stable isotope ratios ($^{15}N/^{14}N$). Soil Biology and Biochemistry. 2004;**36**:1769-1774

[4] Koehler HH. Mesostigmata (Gamasina, Uropodina), efficient predators in agroecosystems. Agriculture, Ecosystems and Environment. 1997;**62**:105-117

[5] Schneider K, Maraun M. Top-down control of soil microarthropods – Evidence from a laboratory experiment. Soil Biology and Biochemistry. 2009;**41**:170-175

[6] Siepel H. Biodiversity of soil microarthropods: The filtering of species. Biodiversity and Conservation. 1996;**5**:251-260

[7] Minor MA, Cianciolo JM. Diversity of soil mites (Acari: Oribatida, Mesostigmata) along a gradient of land use types in New York. Applied Soil Ecology. 2007;**35**:140-153

[8] Erdmann G, Scheu S, Maraun M. Regional factors rather than forest type drive the community structure of soil living oribatid mites (Acari, Oribatida). Experimental & Applied Acarology. 2012;**57**:157-169

[9] Hågvar S, Abrahamsen G. Eksperimentelle forsuringsforsøk i skog 5. Jordbunnszoologiske undersøkelser. SNSF-project; Internal Report 32/77; 1977. pp. 1-47. (In Norwegian, English summary)

[10] Hågvar S, Abrahamsen G. Colonisation by Enchytraeidae, Collembola and Acari in sterile soil samples with adjusted pH levels. Oikos. 1980;**34**:245-258

[11] Leinaas HP. Activity of Arthropoda in snow within a coniferous forest, with special reference to Collembola. Holarctic Ecology. 1981;**4**:127-138

[12] Hågvar S, Kjøndal BR. Succession, diversity and feeding habits of microarthropods in decomposing birch leaves. Pedobiologia. 1981;**22**:385-408

[13] Hågvar S, Kjøndal BR. Effects of artificial acid rain on the microarthropod fauna in decomposing birch leaves. Pedobiologia. 1981;**22**:409-422

[14] Hågvar S, Amundsen T. Effects of liming and artificial acid rain on the mite (Acari) fauna in coniferous forest. Oikos. 1981;**37**:7-20

[15] Hågvar S. Six common mite species (Acari) in Norwegian coniferous forest soils: Relations to vegetation types and soil characteristics. Pedobiologia. 1984;**27**:355-364

[16] Hågvar S. Why do collemboles and mites react to changes in soil acidity? Entomologiske Meddelelser. 1987;**55**:115-119

[17] Hågvar S. Decomposition studies in an easily-constructed microcosm: Effects of microarthropods and varying soil pH. Pedobiologia. 1988;**31**:293-303

[18] Hågvar S. Reactions to soil acidification in microarthropods: Is competition a key factor? Biology and Fertility of Soils. 1990;**9**:178-181

[19] Hågvar S. Succession of microarthropods in decomposing spruce needles: A litter bag study over ten years. In: Veeresh GK, Rajagopal D, Viraktamath CA, editors. Advances in Management and Conservation of Soil Fauna. Proceedings from the 10th International Soil Zoology Colloquium, Bangalore 7-13 August 1988. New Delhi, India: Oxford & IBH Publishing Co. PVT. LTD; 1991. pp. 485-489

[20] Hågvar S. Soil animals and soil acidity. In: Abrahamsen G, Stuanes AO, Tveite B, editors. Long-Term Experiments with Acid Rain in Norwegian Forest Ecosystems. Ecological Studies. Vol. 104. Berlin: Springer-Verlag; 1993. pp. 101-121

[21] Hågvar S. Instability in small, isolated microarthropod communities. Polskie Pismo Entomologiczne. 1995;**64**:123-133

[22] Hågvar S. Mites (Acari) developing inside decomposing spruce needles: Biology and effect on decomposition rate. Pedobiologia. 1998;**42**:358-377

[23] Hågvar S, Hågvar EB. Invertebrate activity under snow in a south-Norwegian spruce forest. Soil Organisms. 2011;**83**(2):187-209

[24] Hågvar S, Steen R. Succession of beetles (genus *Cis*) and oribatid mites (genus *Carabodes*) in dead sporocarps of the red-banded polypore fungus *Fomitopsis pinicola*. Scandinavian Journal of Forest Research. 2013;**28**:436-444. DOI: 10.1080/02827581.2012.755562

[25] Hågvar S, Amundsen T, Økland B. Mites of the genus *Carabodes* (Acari, Oribatida) in Norwegian coniferous forests: Occurrence in different soils, vegetation types and polypore hosts. Scandinavian Journal of Forest Research. 2014;**29**:629-638. DOI: 10.1080/02827581.2014.965195

[26] Hågvar S. From litter to humus in a Norwegian spruce forest: Long-term studies on the decomposition of needles and cones. Forests. 2016;7:186. DOI: 10.3390/f7090186

[27] Hågvar S. Collembola in Norwegian coniferous forest soils I. Relations to plant communities and soil fertility. Pedobiologia. 1982;**24**:255-296

[28] Huhta V, Hyvönen R, Kaasalainen P, Koskenniemi A, Muona J, Mäkelä I, et al. Soil fauna of Finnish coniferous forests. Annales Zoologici Fennici. 1986;**23**:345-360

[29] Siira-Pietikäinen A, Penttinen R, Huhta V. Oribatid mites (Acari: Oribatida) in boreal forest floor and decaying wood. Pedobiologia. 2008;**52**:111-118

[30] Persson T, Bååth E, Clarholm M, Lundkvist H, Söderström BE, Sohlenius B. Trophic structure, biomass dynamics and carbon metabolism of soil organisms in a scots pine forest. Ecological Bulletins. 1980;**32**:419-459

[31] Karppinen E. Ecological and transect survey studies on Finnish Camisiids. Annales Zoologici Societatis Zoologicae Botanicae Fennicae "Vanamo". 1955;**17**(2):1-80

[32] Drift JVD. Analysis of the animal community in a beech forest floor. Tijdschrift voor Entomologie. 1951;**94**:1-168

[33] Karppinen E. Über die Oribatiden (Acar.) der Finnischen Waldboden. Annales Zoologici Societatis Zoologicae Botanicae Fennicae "Vanamo". 1958;**19**(1):1-43

[34] Lebrun P. Écologie et biocénotique de quelques peuplements d'arthropodes édaphiques. Mémoires Institut Royal des Scienses Naturelles de Belgique. 1971;**165**:1-203

[35] Forsslund KH. Studier över det lägre djurlivet i nordsvensk skogsmark. Meddelanden från statens skogforsöksanstalt. 1944;**34**(1):1-1283

[36] Persson T. Markarthropodernas abundans, biomassa och respiration i ett äldre tallbestånd på Ivantjärnsheden, Gästrikland - en första studie. Swedish Coniferous Forest Project, Internal Report; 1975. vol. 31. pp. 1-35

[37] Osler GHR, Cole L, Keith AM. Changes in oribatid mite community structure associated with the succession from heather (*Calluna vulgaris*) moorland to birch (*Betula pubescens*) woodland. Pedobiologia. 2006;**50**(4):323-330

[38] Huhta V, Hyvönen R, Koskenniemi A, Vilkamaa P. Role of pH in the effect of fertilization on Nematoda, Oligochaeta and microarthropods. In: Lebrun P, editor. New Trends in Soil Biology. Proceedings of the VIII. Int. Colloq. of Soil Zoology, Louvain-la-Neuve (Belgium). 1983. pp. 61-73

[39] Bååth E, Berg B, Lohm U, Lundgren B, Lundkvist H, Rosswall T, et al. Effects of experimental acidification and liming on soil organisms and decomposition in a scots pine forest. Pedobiologia. 1980;**20**:85-100

[40] Bonnet L. Caractères généraux des populations thécamoebiennes endogées. Pedobiologia. 1961;**1**:6-24

[41] Stachurska-Hagen T. Acidification experiments in conifer forest. 8. Effects of acidification and liming on some soil animals: Protozoa, Rotifera and Nematoda. SNSF-project. 1980;Internal Report 74/80. pp. 1-23

[42] Abrahamsen G. The influence of temperature and soil moisture on the population density of *Cognettia sphagnetorum* (Oligochaeta: Enchytraeidae) in cultures with homogenized raw humus. Pedobiologia. 1971;**11**:417-424

[43] Abrahamsen G. Ecological study of Enchytraeidae (Oligochaeta) in Norwegian coniferous forest soils. Pedobiologia. 1972;**12**:26-82

[44] Abrahamsen G. Effects of lime and artificial acid rain on the enchytraeid (Oligochaeta) fauna in coniferous forest. Holarctic Ecology. 1983;**6**:247-254

[45] Satchell JE. Some aspects of earthworm ecology. In: Kevan DKMCE, editor. Soil Zoology. London: Butterworths; 1955. pp. 180-201

[46] Abrahamsen G. Ecological study of Lumbricidae (Oligochaeta) in Norwegian coniferous forest soils. Pedobiologia. 1972;**12**:267-281

[47] Nordström S, Rundgren S. Environmental factors and lumbricid associations in southern Sweden. Pedobiologia. 1974;**14**:1-27

[48] Christiansen K. Competition between collembolan species in culture jars. Revue d'écologie et de biologie du sol. 1967;**4**:439-462

[49] Nygard J, Solberg J. Laboratory study on competition between four soil-living species of Collembola. Thesis, University of Oslo (in Norwegian); 1985

[50] Anderson JM. Competition between two unrelated species of soil cryptostigmata (Acari) in experimental microcosms. Journal of Animal Ecology. 1978;**47**:787-803

[51] Anderson JM. Succession, diversity and trophic relationships of some soil animals in decomposing leaf litter. Journal of Animal Ecology. 1975;**44**:475-495

[52] Mignolet R, Lebrun P. Colonisation par les microarthropodes du sol de cinq types de litière en decomposition. In: Vanek J, editor. Progress in Soil Zoology. The Hague, The Netherlands: Dr. W. Junk, B.V; 1975. pp. 261-281

[53] Mignolet R. Écologie des microarthropodes associés à la decomposition des litières de forêt. Thesis, Univ. Catholique de Louvain. 1975. 136 pp

[54] Franklin E, Hayek T, Fagundes EP, Silva LL. Oribatid mite (Acari: Oribatida) contribution to decomposition dynamic of leaf litter in primary forest, second growth, and polyculture in the Central Amazon. Brazilian Journal of Biology. 2004;**64**. DOI: 10.1590/S1519-69842004000100008

[55] Fujii S, Takeda H. Succession of soil microarthropod communities during the aboveground and belowground litter decomposition processes. Soil Biology and Biochemistry. 2017;**110**:95-102

[56] Pande YD, Berthet P. Studies on the food and feeding habits of soil Oribatei in a black pine plantation. Oecologia. 1973;**12**:413-426

[57] Ponge J-F. Succession of fungi and fauna during decomposition of needles in a small area of scots pine litter. Plant and Soil. 1991;**138**(1):99-113

[58] Coulianos C-C, Johnels AG. Note on the subnivean environment of small mammals. Arkiv för Zoologi. 1962;**Ser.2**:363-370

[59] Gulvik M. Mites (Acari) as indicators of soil biodiversity and land use monitoring: A review. Polish Journal of Ecology. 2007;**55**:415-440

[60] Hågvar S. The relevance of the Rio-convention on biodiversity to conserving the biodiversity of soils. Applied Soil Ecology. 1998;**9**:1-7

[61] Hågvar S. The Huge Biodiversity of Soils: Can it be Saved? Agricultural University of Iceland, Hvanneyri. AUI Publication. 2006;**9**:61-66

Chapter 7

Invasive Mite Species in the Americas: Bioecology and Impact

Carlos Vásquez and Yelitza Colmenárez

Abstract

Invasive species represent one of the most relevant threats for biodiversity in many ecosystems, mainly in those so-called agroecosystems due to which they exhibit reduced biodiversity and simplified trophic interactions. These two factors make many niches unoccupied, thus increasing the risk that invasive species especially arthropod pests occupy these niches or compete with native species. In spite of potential impact of invasive species, our understanding of their ecological consequences is developing slowly. In the last years, more attention is being paid on phytophagous mites because several noneconomic species have become severe pests on many crops as a consequence of irrational use of agrochemicals. Also, due to the small size of the mites, they can be transported throughout the world and established in new areas where favorable conditions and the absence of efficient natural enemies favor their development. Thus, phytophagous mites are feasible to become invasive species since they are able to provoke severe damage to plants. Since 2004, *Steneotarsonemus spinki*, *Schizotetranychus hindustanicus,* and *Raoiella indica* have been introduced in the Neotropical region. Information about pest status, seasonal trends, and natural enemies in invaded areas is provided for these species.

Keywords: phytophagous mites, invasive species, Neotropical region, *Steneotarsonemus spinki*, *Schizotetranychus hindustanicus*, *Raoiella indica*

1. Introduction

Natural environments are continuously submitted to severe transformations, including movement of species beyond the limits of their native geographic ranges into areas where they do not naturally occur and where they can inflict substantial changes [1]. Thus, considering changes inflicted by alien species to the properties of an ecosystem, an increasing number of studies that consider the environmental impacts have been published [2]. However, according to Ricciardi [3], a predictive understanding of the ecological impacts of invasive species has developed slowly, owing largely to an apparent lacking of clearly defined hypotheses and of a broad theoretical framework. In this regard, confusion about terminology used for the designation of nonindigenous species, which alternatively have been called "exotic," "introduced," "invasive," and "naturalized," is particularly acute, which leads to confusion about ecological concepts [4].

Another term needing delimitation of definition is referred to concept of impact. On an ecological basis, an impact is defined as a measurable change to the properties of an ecosystem by a nonnative species, which is considered to provoke a positive or negative impact simply by becoming integrated into the system [3].

Various studies have shown that nonnative species can promote extinction of native species, and also they can provoke changes in genetic composition of native populations, behavior patterns, species richness and abundance, phylogenetic and taxonomic diversity, etc.

Finally, when considering invasive species in an agricultural ambit, it is strongly recommended to define invasion threat, which is conceptualized as the likelihood of a particular pest or pathogen arriving in a new location as well as the establishment likelihood considered as the chances of those pests or pathogens to establish in a new location [5].

Plant and animal species have been transported by humans for millennia; even a well-defined period in biological invasions dates as far back as 1500 AD, a period associated with the birth of colonialism and the start of radical changes in patterns of human demography, agriculture, trade, and industry [6]. However, more recently, increasing globalization and world trade have augmented the possibility of arrival of invasive species to geographic regions in which they were previously absent, making necessary to quantify impact of invasive species and develop effective biosecurity policy [4].

Since the end of the twentieth century, more attention is being paid on phytophagous mites because several noneconomic species have become severe pests on many crops as a consequence of irrational use of agrichemicals. Also, due to the small size of the mites, they can be transported throughout the world and set up in new areas, in which favorable conditions and the lack of efficient natural enemies favor their development, resulting in economic losses [7]. There are various examples of introductions of phytophagous mites in new areas such as the cassava green mite [*Mononychellus tanajoa* (Bondar, 1938)], the coconut mite [*Aceria guerreronis* (Keifer, 1965)], and the tomato spider mite [*Tetranychus evansi* Baker and Pritchard, 1960] [7, 8]. Both *M. tanajoa* and *A. guerreronis* were introduced into Africa, while *T. evansi* has been introduced in Africa in the Mediterranean Basin.

Similarly, some phytophagous mite species have been introduced in the Neotropical region, i.e., *Steneotarsonemus spinki* (Smiley, 1967), *Schizotetranychus hindustanicus* (Hirst, 1924), and *Raoiella indica* (Hirst, 1924). In the present review, information about recent phytophagous mite on pest status, seasonal trends, and natural enemies in invaded areas is provided for these species. Because invasive species may evolve during the invasion process, comparison of behavior, and damage and management options between native and invaded areas for these species will be useful for understanding the invader's success and their ability to colonize new regions.

2. Some concepts related to invasive species

The introduction of species beyond their native range as a direct or indirect result of human action causes changes in the ecosystems to which they are introduced [9]. Moreover, these biological invasions are causing tremendous damages to ecosystems and economic activities [10]. Many important terms related to the invasion ecology, such as "invasive," "weed," or "transient," can be susceptible to subjective interpretation, consequently causing a lack of consensus about terms used to define nonindigenous species [4]. Thus, some terms such as "noxious" and "nuisance" are generally used to indicate direct or indirect adverse effects on humans; however, according to Colautti and MacIsaac [4], interactions have three important implications:

a. It is necessary to define if invasive species cause aesthetical displeasing effects and are vectors for serious human diseases.

b. Species might be erroneously considered as an environmental threat (or weedy, invasive, etc.) in areas where they have little or no impact only based on these species had been identified as a disturbance elsewhere, disregarding thus the ecological phenomenon.

c. A particular species can have both beneficial and detrimental effects. *Harmonia axyridis* (Pallas, 1773) (Coleoptera, Coccinellidae) is a well-known predator widely used as a classical biological control agent of aphids around the world; however, this species has provoked the displacement of native aphid species in Brazil and other South American countries [11].

Some definitions of invasive species are the following:

a. An alien species whose introduction does or is likely to cause economic or environmental harm or harm to human health.

b. A species that is nonnative to the ecosystem under consideration and whose introduction causes or is likely to cause economic or environmental harm or harm to human health

Invasion consists of a series of steps that an organism must undergo to become a successful invader, and thus, it can inflict an ecological damage [12]. According to Beck et al. [13], successful invasion is preceded by the following stages:

a. *Large-scale geographical barriers*: species from a geographical area is supposed to overcome a geographical barrier (mountain range, ocean, or other physical barriers) to arrive in a new area where it does not previously occur (so-called alien species, nonnative species). This movement is often mediated by human activities, either deliberately or unintentionally.

b. *Survival barriers*: these often refer to environmental barriers such as environmental conditions that let the nonnative organism to survive and develop in its new location. Other survival barriers may include host plants, competitor organisms, predators, and pathogens.

c. *Establishment barriers*: depending on the survival abilities of the alien species, it will be able to form a self-sustaining population and does not need a re-introduction to support a population base.

d. *Dispersal and spread barriers*: once established, alien species must disperse and spread relatively fast from their site of establishment. However, this movement or spread alone does not necessarily make this nonnative species an invasive species.

3. Mite invasions in the Neotropical region

Although bioinvasions have occurred for many years, most documented cases have been reported in recent decades and even invasive or adventive mites have gained attention only in the last few years when they have been the target of research to determine their potential distribution [7]. After colonization, it is fundamental to determine the intrinsic properties of the invasive population, the genetic structure of populations, and the response to environmental factors to developing strategies and policies of management [7].

In the Neotropical region, several mite species have recently invaded the agricultural landscapes in Latin America, for example, the citrus Hindu mite, *S. hindustanicus* (Tetranychidae), the rice mite, *S. spinki* (Tarsonemidae), and the red palm mite, *R. indica* (Tenuipalpidae).

3.1 *Schizotetranychus hindustanicus*

The genus *Schizotetranychus* includes 114 species; however, information about economic importance of most of the species is still scarce [14, 15] (**Table 1**). Most of the species occur in Asia and CIS, and only 20 species (17.5%) are in the Neotropical region, including *S. hindustanicus*.

The citrus Hindu mite, *S. hindustanicus*, was originally described from citrus from southern India (Hirst, 1924), and its occurrence had been reported in this country for almost 80 years; however, in 2005, this species was surprisingly found in the northwestern Venezuela [16] and soon after in Colombia and Brazil [17].

S. hindustanicus had only been reported on four host plant species in India (see **Table 1**); however, posteriorly, it was found on *Acacia* sp., *Melia azedarach* L. and various *citrus* species (**Figure 1**). Symptoms of mite feeding first appear on the upper leaf surface, along the main rib, later extending to the entire leaf; while when feeding on fruits the females webs over concavities or depressions in the rind; attacked fruits become uniformly silvered and hard under severe infestation [18]. Návia and Marsaro [17] reported that although this damage by mite feeding is supposed to affect the commercial value of infested fruits, nothing has been published about the resulting economic impact.

In Venezuela, *S. hindustanicus* has been observed forming colonies in several *citrus* species and/or varieties such as *C. latifolia* (Tanaka ex Yu. Tanaka), *C. aurantifolia* (Chistm) *C. reticulata* Blanco, *C. limon* (L.), and *C. sinensis* (L.) Osbeck [19, 20].

In Colombia, this tetranychid mite species was first reported in the northern coast in Dibulla (Guajira) and Magdalena [21]. After that, ICA (Agropecuary Colombian Institute) carried out samplings in departments of Atlantico, Bolivar, Guajira, Magdalena, and Vichada as shown in **Table 2**. Similarly, presence of circular whitish spots on leaves and fruits of "tahiti" and "galeguinho" lemon trees in urban areas of Roraima (Brazil) is alerted to the Brazilian plant protection authorities as this country is the largest *citrus* producer [17]. According to these authors, dispersion of *S. hindustanicus* could cause high economic impact and/or commercial restrictions due to sanitary.

Since these tetranychid mite species can be the pest on *citrus* spp., some studies have been carried out in Venezuela. Niedstaedt and Marcano [15] observed the effect that the developmental time of *S. hindustanicus* varied from 30.12 to 31.10 days on sweet orange or Persian lime, respectively, at 25°C. Additionally, population studies on Persian lime, lemon, sweet orange, and tangerine showed that number of individuals was relatively low in two peaks: the first peak during June 2005 with 24.17, 21.67, and 12 individuals was observed on tangerine, sweet orange, and Persian lime, respectively, while the second peak with lightly higher number of mites developed during April 2006 was observed with 69.17, 31.2, and 20.2 mites on sweet orange, tangerine, and Persian lime, respectively [22].

Field observations on different citrus species have demonstrated that *S. hindustanicus* can colonize which seems to be verifying the entire canopy so far economic impact have not been evaluated in the neotropical areas. There are some studies on this genus, associated mainly with grasses such as rice and bamboo and some fruit trees [23, 24].

Species	Host plants	Distribution
S. agropyron (Tuttle and Baker, 1976)	*Agropyron desertorum* (Fisch. ex Link) Schult.	The USA
S. alni (Beglyarov and Mitrofanov, 1973)	*Alnus* sp.	CIS
S. andropogoni (Hirst, 1926)	*Andropogon annulatus* Forssk., *Chloris incomplete* Roth, *Dichanthium annulatum* (Forssk.) Stapf, *Oryza sativa* L.	CIS, India, Mexico, Pakistan, and Thailand
S. approximatus (Ehara, 1988)	*Bambusa vulgaris* Schrad. ex J.C. Wendl., *Thyrsostachys siamensis Gamble*	Malaysia
S. arcuatus (Meyer, 1974)	*Euclea crispa* (Thunb.) Gürke	South Africa
S. asparagi (Oudemans, 1928)	*Acacia horrida* (L.) Willd., *A. longifolia* (Andrews) Willd., *Ananas* sp., *Aspalathus* sp., *Asparagus* sp., *A. africanus* Lam., *A. officinalis* L., *A. plumosus* Baker, *A. setaceus* (Kunth) Jessop, *A. sprengeri* Regel, *A. suaveolens* Burch., *Protasparagus capensis* (L.) Oberm., *P. compactus* (T.M. Salter) Oberm., *P. laricinus* (Burch.) Oberm., *Pteridophyta*	Australia, Germany, Hawaii, Israel, Morocco, Portugal, Puerto Rico, South Africa, the Netherlands, and the USA
S. australis (Gutierrez, 1968)	*Mundulea pungens* R. Vig., *Tephrosia striata* Ecklon & Zeyher ex Steudel	Madagascar
S. avetjanae (Bagdasarian, 1954)	*Spiraea* sp.	CIS
S. baltazari (Rimando, 1962)	*Citrus grandis* (L.) Osbeck, *C. madurensis* Lour., *C. medica* L., *C. sinensis* (L.) Osbeck	Burma, China, Hong Kong, India, Philippines, Taiwan, and Thailand
S. bambusae (Reck, 1941)	*Arundinaria* sp. *Phyllostachys* sp., *P. nigra (Lodd. ex Lindl.) Munro, P. reticulata* (Rupr.) K. Koch	CIS, China, Hainan Island, Japan, and Korea
S. beckeri (Wainstein, 1958)	*Calamagrostis* sp., *Dactylis* sp., *Helictotrichon* sp.	CIS
S. bhandhufalcki (Ehara and Wongsiri, 1975)	*Cajanus cajan* (L.) Huth, *Cassia* sp., *C. siamea* Lam., *Colocasia esculenta* (L.) Schott, *Pterocarpus macrocarpus* Kurz	Thailand
S. boutelouae (Tuttle and Baker, 1968)	*Bouteloua rothrockii* Vasey, *Commelina dianthifolia* L., *Stipa eminens* Cav.	Mexico and the USA
S. brachypodii (Livshits and Mitrofanov, 1968)	*Brachypodium silvaticum* (Huds.) P. Beauv.	CIS
S. brevisetosus (Ehara, 1989)	*Quercus* sp., *Q. glauca* Thunb.	Japan
S. cajani (Gupta, 1976)	*C. cajan*	India

Species	Host plants	Distribution
S. camur (Pritchard and Baker, 1955)	*Arundinaria* sp.	USA
S. celarius (Banks, 1917)	*Arundinaria hindsii* Munro, *Bambusa* sp., *Ficus stipulata* Thunb., *Miscanthus sinensis* Andersson, *Oryza* sp., *Phyllostachys* sp., *P. makinoi* Hayata, *P. nigra* (Lodd. ex Lindl.) Munro, *P. reticulata* (Rupr.) K. Koch	Australia, China, France, Hawaii, Hong Kong, Japan, Korea, Okinawa Island, Taiwan, the Netherlands, and the USA
S. celtidis (Tuttle and Baker, 1968)	*Celtis reticulata* Torr., *Leptochloa uninervia* (J. Presl) Hitchc. & Chase, *Sporobolus flexuosus* (Thurb. ex Vasey) Rydb., *Tridens pulchellus* (Kunth) Hitchc.	Mexico and the USA
S. cercidiphylli (Ehara, 1973)	*Cercidiphyllum japonicum* Siebold & Zucc.	Japan
S. chiangmaiensis (Ehara and Wongsiri, 1975)	*Calotropis gigantea* (L.) W.T. Aiton	Thailand
S. colocasiae (Ehara, 1988)	*Colocasia* sp.	Malaysia
S. cornus (Pritchard and Baker, 1955)	*Dysoxylum spectabile* Hook. f., *Elaeocarpus dentatus* (J.R. Forst. and G. Forst.) Vahl	New Zealand
S. cynodontis (McGregor, 1950)	*Agrostis* sp., *Cynodon dactylon* (L.) Pers.	USA
S. dalbergia (Meyer, 1974)	*Dalbergia melanoxylon* Guill. & Perr.	Zimbabwe
S. denmarki (Baker and Tuttle, 1994)	Poaceae	USA
S. echinulatus (Mitrafanov, 1978)	*Spiraea* sp.	CIS
S. elongates (Wang and Cui, 1991)	Bambusaceae	China
S. elymus (McGregor, 1950)	*Agropyron* sp., *Agrostis* sp., *Aristida adscensionis* L., *Bouteloua hirsuta* Lag., *C. dactylon*, *Distichlis stricta* (Torr.) Rydb., *Elymus* sp., *E. trachycaulus* (Link) Gould, *Hordeum* sp., *Malva parviflora* L., *Panicum obtusum* Kunth, *P. scribnerianum* Nash, *Stipa ichu* (Ruiz & Pav.) Kunth, *Tridens pulchellus* (Kunth) Hitchc., *Triticum aestivum* L., *Typha latifolia* L., *Vicia pulchella* Kunth	Mexico and the USA
S. emeiensis (Wang, 1983)	Bambusaceae	China
S. eremophilus (McGregor, 1950)	*Aristida adscensionis* L., *A. glabrata* (Vasey) Hitchc., *Bothriochloa saccharoides* (Sw.) Rydb., *Bouteloua* sp., *B. barbata* Lag., *C. dactylon*, *Distichlis stricta* (Torr.) Rydb., *Lycurus phleoides* Kunth, *Tridens pulchellus* (Kunth) Hitchc.	Mexico and the USA
S. euphorbiae (Livshits and Mitrofanov, 1968)	*Euphorbia amygdaloides* L.	CIS

Species	Host plants	Distribution
S. fauveli (Gutierrez, 1978)	*Ficus edulis* Burm. f., *F. fraseri* Miq., *F. habrophylla* G. Benn. & Seem.	New Caledonia
S. filifolius (Meyer, 1974)	*Aster filifolius* Vent.	South Africa
S. floresi (Rimando, 1972)	*Arundo formosana* Hack., *Bambusa* sp., *B. spinosa* Roxb.	Philippines, Taiwan
S. fluvialis (McGregor, 1928)	*A. adscensionis, C. cajan, Epicampes rigens* Benth., *Muhlenbergia rigens* (Benth.) Hitchc.	India, USA
S. freitezi (Ochoa, Gray and von Lind., 1990)	*O. sativa*	Costa Rica
S. gahniae (Davis, 1969)	*Gahnia aspera* Spreng.	Australia
S. garmani (Pritchard and Baker, 1955)	*Acer* sp., *Populus tremula* L., *Quercus* sp., *Q. robur* L., *Salix* sp., *S. caprea* L., *S. humilis* Marshall, *S. petiolaris* Sm., *S. tristis* Aiton	Iran, Poland, Switzerland, and the USA
S. gausus (Baker and Pritchard, 1960)	Unknown	Zaire
S. glabrisetus (Ugarov and Nikolskii, 1937)	Poaceae	CIS
S. graminicola (Goux, 1949)	*Molinia caerulea* Milk.	France and the Netherlands
S. guatemalae-novae (Stoll, 1886)	*Cassia nictitans* L.	Guatemala
S. halimodendri (Waistein, 1958)	*Halimodendron halodendron* (Pall.) Druce	CIS
S. hilariae (Tuttle and Baker, 1968)	*A. adscensionis, Hilaria rigida* (Thurb.) Benth. ex Scribn.	USA
S. hindustanicus (Hirst, 1924)	*Azadirachta indica* A. Juss., *Citrus* sp., *Cocos nucifera* L., *Sorghum bicolor* (L.) Moench	India
S. ibericus (Reck, 1947)	*Quercus* sp.	CIS
S. imperatae (Wang, 1983)	*Imperata* sp.	China
S. jachontovi (Reck, 1953)	*Quercus* sp.	CIS
S. kaspari (Manson, 1967)	*Calopogonium mucunoides* Desv., *Cordyline kaspar* W.R.B. Oliv.	Hainan Island and New Zealand
S. kochummeni (Ehara, 1988)	Bambusaceae	Malaysia

Species	Host plants	Distribution
S. laevidorsatus (Ehara, 1988)	Bambusaceae, *Gigantochloa levis* (Blanco) Merr.	Malaysia
S. lanyuensis (Tseng, 1975)	*Unknown*	Taiwan
S. lechrius (Rimando, 1962)	*Cassia siamea* Lam., *Citrus* sp., *C. esculenta, Glycine max* (L.) Merr., *Pterocarpus indicus* Willd., *P. vidalianus* Rolfe	Indonesia, Philippines, and Taiwan
S. lespedeza (Beglyarov and Mitrofanov, 1973)	*Bauhinia* sp., *Desmodium* sp., *Lespedeza* sp., *L. bicolor*	CIS, China, Japan, Korea, and Taiwan
S. levinensis (Manson, 1967)	Poaceae	New Zealand
S. longirostrus (Feres and Flechtmann, 1995)	*Bambusa* sp.	Brazil
S. longus (Saito, 1990)	*Sasa senanensis* (Franch. & Sav.) Rehder	Japan
S. luculentus (Tseng, 1990)	*Diospyros* sp.	Taiwan
S. lushanensis (Dongsheng, 1994)	*Cinnamomum camphora* Meisn.	China
S. lycurus (Tuttle and Baker, 1964)	*Leersia oryzoides* (L.) Sw., *Lycurus phleoides* Kunth, *Setaria macrostachya* Kunth	Mexico and the USA
S. malayanus (Ehara, 1988)	*Manihot* sp.	Indonesia and Malaysia
S. malkovskii (Waistein, 1956)	*Calamagrostis* sp.	CIS
S. mansoni (Gupta, 1980)	*Oryza* sp.	India
S. minutus (Wang, 1985)	Bambusaceae	China
S. miscanthi (Saito, 1990)	*Miscanthus* sp., *M. sinensis* Andersson	Japan
S. miyatahus (Meyer, 1974)	*Pterocarpus rotundifolius* (Sond.) Druce	South Africa
S. montanae (Tuttle and Baker, 1968)	*Muhlenbergia montana* (Nutt.) Hitchc., *Pappophorum mucronulatum* Nees	Mexico and the USA
S. nanjingensis (Ma and Yuan, 1980)	*Phyllostachys* sp.	China
S. nesbitti (Meyer, 1965)	Poaceae	South Africa
S. nugax (Pritchard and Baker, 1955)	*Hilaria mutica* (Buckley) Benth., Poaceae	Mexico and the USA

Species	Host plants	Distribution
S. oryzae (Rossi de Simons, 1966)	*O. sativa, Panicum maximum* Jacq.	Argentina, Brazil, Colombia, Surinam, and Venezuela
S. oudemansi (Reck, 1948)	*Vaccinium uliginosum* L.	CIS
S. paezi (Alvarado and Freitez, 1976)	*O. sativa, P. maximum*	Colombia and Venezuela
S. papillatus (Flechtmann, 1995)	Bambusaceae	Brazil
S. paraelymus (Feres and Flechtmann, 1995)	*Bambusa* sp.	Brazil
S. parasemus (Pritchard and Baker, 1955)	*C. dactylon, Dactylis glomerata* L., *Distichlis spicata* (L.) Greene, *D. stricta* (Torr.) Rydb., *Paspalum notatum* Flüggé, *Vitis* sp.	Brazil, Colombia, Poland, and the USA
S. pennamontanus (Meyer, 1987)	*Lebeckia linearifolia* E. Mey.	Namibia
S. prosopis (Tuttle, Baker and Abbatiello, 1976)	*Prosopis juliflora* (Sw.) DC.	Mexico
S. protectus (Meyer, 1975)	*Cliffortia linearifolia* Eckl. & Zeyh., *C. repens* Schltr.	South Africa
S. pseudolycurus Ochoa, (Gray and von Lind., 1990)	*O. sativa*	Costa Rica and Panamá
S. recki (Ehara, 1957)	*S. senanensis*	Japan
S. reticulatus (Baker and Pritchard, 19600	*Grewia* sp.	Comoro Island and Zaire
S. rhodanus (Baker and Pritchard, 1960)	Unknown	Zaire
S. rhynosperus (Flechtmann and Baker, 1970)	*Rhynchospora* sp.	Argentina and Brazil
S. russeus (Davis, 1969)	*Lomandra multiflora* Britten	Australia
S. saba-sulchani (Reck, 1956)	*C. dactylon*	CIS
S. saccharum (Flechtmann and Baker, 1975)	*Saccharum officinarum* L.	Brazil

Species	Host plants	Distribution
S. sacrales (Baker and Pritchard, 1960)	Fabaceae	Zaire
S. sagatus (Davis, 1969)	*Themeda australis* (R. Br.) Stapf	Australia
S. saitoi (Ehara, 1968)	*B. vulgaris*	Malaysia
S. sayedi (Attiah, 1967)	*Ficus carica* L.	Egypt
S. schizopus (Zacher, 1913)	*Populus* sp., *P. tremula, Salix* sp., *S. alba* L., *S. aurita* L., *S. balsamifera* (Hook.) Barratt ex Andersson, *S. bicolor* Willd., *S. caprea* L., *S. daphnoides* Vill., *S. elegantissima* K. Koch, *S. fragilis* L., *S. nigra* Marshall, *S. purpurea* L., *S. subfragilis* Andersson, *S. viminalis* L., *Vaccinium uliginosum* L.	CIS, China, Germany, Hungary, Japan, Poland, Switzerland, the Netherlands, the UK, and the USA
S. setariae (Meyer, 1987)	*Setaria sphacelata* (Schumach.) Stapf & C.E. Hubb. ex M.B. Moss	South Africa
S. smirnovi (Waistein, 1954)	*Juglans regia* L., *Malus domestica* (Suckow) Borkh., *Morus alba* L., *Prunus armeniaca* L.	CIS
S. spicules (Baker and Pritchard, 1960)	*Citrus* sp., *Murraya koenigii* (L.) Spreng.	India and Kenya
S. spiraefolia (Garman, 1940)	*C. cajan*, *S. officinarum*, *Spiraea* sp., *S. latifolia* (Aiton) Borkh., *S. pubescens* Turcz., *S. salicifolia* L., *S. trilobata* L.	China, India, Poland, and the USA
S. taquarae (Paschoal, 1971)	*B. vulgaris*	Brazil
S. tbilisiensis (Reck, 1959)	*Agrostemma githago* L., *Bromus* sp., *Elytrigia repens* (L.) Desv. ex Nevski, *Marrubium* sp.	CIS
S. tephrosiae (Gutierrez, 1968)	*Balanites pedicellaris* Mildbr. & Schltr., *Eriobotrya japonica* (Thunb.) Lindl., *Mikania cordata* (Burm. f.) B.L. Rob., *Mundulea pungens* R. Vig., *M. sericea* Hook. & Arn., *Tephrosia striata* Ecklon & Zeyher ex Steudel	India, Madagascar, and South Africa
S. textor (Waisntein, 1954)	*Elaeagnus angustifolia* L., *Lonicera* sp.	CIS
S. triquetrus (Meyer, 1987)	*Pentzia incana* (Thunb.) Kuntze	South Africa
S. tuberculatus (Ugarov and Nikolskii, 1937)	*Morus* sp.	CIS
S. tumidus (Wang, 1981)	*Melia radula* (nonvalid name)[1]	China
S. tuminicus (Ma and Yuan, 1982)	*Bridelia monoica* (Lour.) Merr.	China
S. tuttlei (Zacher, Gomaa and El-Enany, 1982)	*Arundo donax* L., *Cuscuta planiflora* Ten., *Mentha pulegium* L., *O. sativa*	Egypt

Species	Host plants	Distribution
S. ugarovi (Wainstein, 1960)	*Alhagi pseudalhagi* (M. Bieb.) Desv. ex B. Keller & Shap.	CIS
S. umtaliensis (Meyer, 1974)	*Acacia* sp.	Zimbabwe
S. undulates (Beer and Lang, 1958)	*Acacia nilotica* (L.) Willd. ex Delile, *Beaucarnea stricta* Lem., *Jasminum grandiflorum* L.	India and Mexico
S. vermiculatus (Ehara and Wongsiri, 1975)	Poaceae	Thailand
S. yoshimekii (Ehara and Wongsiri, 1975)	*O. sativa*	China and Thailand
S. youngi (Tseng, 1975)	*Citrus medica* L., *C. paradisi* Macfad.	Taiwan
S. zhangi (Wang and Cui, 1992)	*Quercus gilliana* Rehder & E.H. Wilson	China
S. zhongdianensis (Wang and Cui, 1992)	*Salix* sp.	China

[1]*Species names that are nonvalid according to the MOBOT.*

Table 1.
Worldwide Schizotetranychus species (from [14]).

Figure 1.
Schizotetranychus hindustanicus colony on citrus leaves (a) and citrus leaves showing characteristic symptoms for S. hindustanicus feeding (courtesy of Dr.). (b) Mario Cermelli and Pedro Morales.

3.2 *Steneotarsonemus spinki*

The rice mite, *S. spinki*, is the origin of southeastern Asia, where it has been reported causing damage to rice crops varying from 30 to 90% in China and 20–60% in Taiwan [25]. Presently, it is considered as a serious pest of rice in Tropical Asia and Caribbean [26]. Other than rice, *S. spinki* is associated to more than 70 plant species including weeds growing near rice fields, such as wild rice:

Department	Municipality	Presence/absence	Host plant
Atlántico	Luruaco	–	
	Polonuevo	–	
	Baraona	–	
La Guajira	Dibulla	+	
Magdalena	Santa Marta	+	
	Santa Ana	–	
	Ciénaga	–	
	Zona Bananera	–	
	El Banco		
	Guamal	+	*C. sinensis* *C. limon*
	San Sebastián	–	
	San Zenón	–	
Vichada	Puerto Carreño	–	

Table 2.
Surveyed localities in the northern Colombia to detect occurrence of S. hindustanicus (from ICA, 2012).

O. latifolia, *C. dactylon* (Poaceae), *Cyperus articulatus* L., *Cyperus iria* L., and *Oxycaryum* sp. (Cyperaceae) [26, 27].

The rice mite feeds on the adaxial surface of leaf sheaths and developing kernels evidenced by brown lesions and consequently reducing photosynthesis and having a negative effect on fertility [26]. Damage also results in sterile grain syndrome, which is characterized by losing and brown discoloration of the flag leaf sheath, twisted panicle neck, and impaired grain development with empty or incompletely filled grains with brown spots and panicles standing erect. The damage to grains showing sterility and malformed curved appearance is referred to as "parrot-beak" [28]. On the other hand, Shikata et al. [29] found for the first time virus-like particles associated with the tarsonemid mites in the rice plants; the spherical virus-like particles were isolated from the rice plants infected with rice ragged stunt, dwarf, black-streaked dwarf, grassy stunt viruses, as well as from the "healthy" plants, which were not inoculated with those viruses, and in addition, the same particles were also found in the dip preparations of the rice tarsonemid mites and eggs.

S. spinki was first reported in North America in 1960 on *Tagosodes orizicolus* (Muir, 1926) in Louisiana, USA. Several years after, the rice mite was found causing damage in rice crops (*O. sativa*) in Cuba in 1997 [30]. Subsequently, this tarsonemid mite spreads over all the Caribbean and Central America: Dominican Republic [31], Costa Rica [32], Haiti [33], Panama [34], Guatemala, Honduras [35], and Mexico [36]. In South America, it has been reported in Colombia [37] and Venezuela [38].

After being introduced in Cuba, outbreaks were registered from 1997 to 1998 when an increase in vain grains of 15–20% and a loss of 2 t/ha were recorded [25]. At the end of 1998, *S. spinki* was also found in Dominican Republic and Haiti, causing about 30% of yield loss; however, less intense damage was verified as compared to Cuba [34].

3.3 *Raoiella indica*

The red palm mite, *R. indica*, is of Asian origin, and it is widely distributed in India, Pakistan, Russia, Iran, Israel, Oman, Pakistan, Egypt, Sudan, and Mauritius [39]. Since 2004, *R. indica* was reported from several Caribbean islands, including

Martinique [40], Saint Lucia and Dominica [41], Guadeloupe and Saint Martin [42], Puerto Rico and Culebra Island [43], and Jamaica [44] (Welbourn, 2007). More recently, it has also been found in Venezuela [45], Colombia [46], and Brazil [47].

Raoiella indica can cause severe damage not only to Arecaceae, especially coconut (*C. nucifera*), but also to Musaceae and other plant families [40, 42, 48] (**Figure 2**). Infested plants exhibit a characteristic "yellowing" as a result of mites feeding on the nutrient-rich layers of the leaves' mesophyll tissues [40]. *Raoiella* species inflict damage by introducing the infrastratum through the stomatal opening to feed on the underlying mesophyll cells [49]. Therefore, the distribution of the stomata on the leaf surface could have a greater influence on the feeding capacity of *R. indica* on the host plant [50], and the severity of the feeding damage by the red palm mite increases in the young plants [43, 51].

After RPM was reported in the New World, little was known about bioecology of this Red palm mite. Regarding host plant, only coconut and *Adonidia merrillii* (Becc.) Becc. had been recorded as host plants to this mite [41, 48]. After RPM occurrence in South America in the coastal Sucre state of Venezuela, Vásquez et al. [52] registered higher population levels of RPM on coconut, banana (*Musa* spp.), ornamental plants, and weeds in the northern Venezuela (**Figure 3** and **Table 3**). These authors observed all RPM stages only on eight arecaceous, one musaceous,

Figure 2.
Coconut and plantain trees showing symptoms of R. indica feeding (A, B) and a colony of the red palm mite on coconut leaves (C).

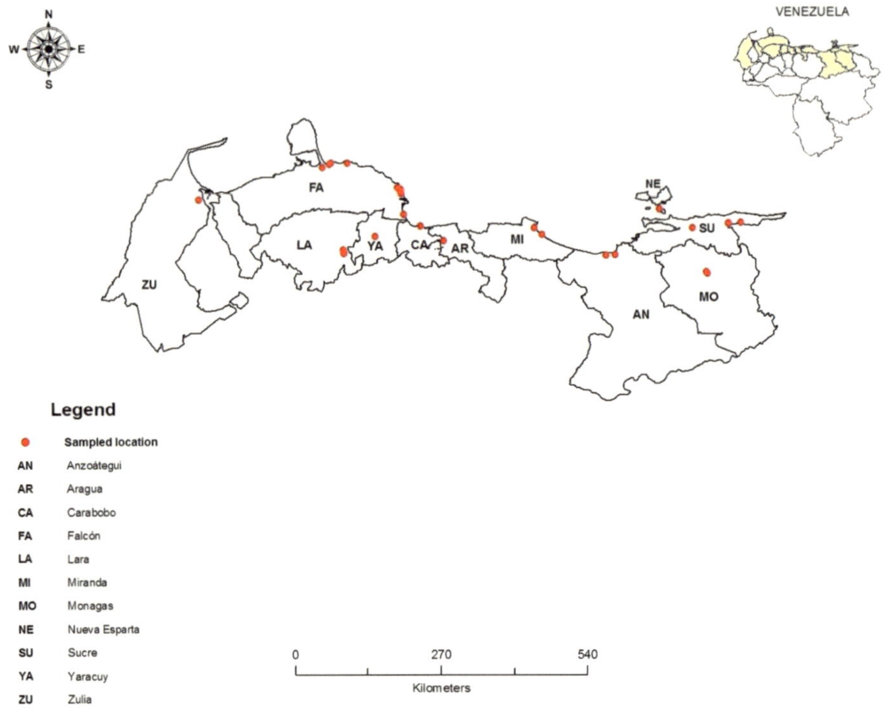

Figure 3.
Distribution of R. indica in Venezuela based on collection records (from 2008 to 2012) [52].

Plant family	Species
Apocynaceae	*Rauvolfia viridis*
Arecaceae	*Adonidia merrillii*
	Cocos nucifera
	Roystonea oleracea
	Pritchardia pacifica
	Ptychosperma macarthurii
	Roystonea regia
	Washingtonia sp.
	Washingtonia robusta
	Phaseolus sp.
Musaceae	*Musa* sp.
Sterculiaceae	*Sterculia* sp.
Strelitziaceae	*Strelitzia* sp.

Table 3.
Plant species onto which Raoiella indica was found in the northern Venezuela (from [52]).

and one streliziaceous species, indicating that the pest developed and reproduced only on these plants, while specimens found on weeds were considered spurious events. Later, the list of host plants increased including 73 species of Arecaceae, six of Musaceae, five of Heliconiaceae, four of Zingiberaceae, and two each of Pandanaceae and Strelitziaceae [53].

Due to the potential impact of *R. indica*, in 2007, the Brazilian Ministry of Agriculture added the RPM to the list of quarantine pests so that an extensive survey was initiated in the state of Roraima, a Brazilian state bordering Venezuela [54].

Only after 2 years, in July 2009, the red palm mite was found in samples of coconut and banana leaves in urban areas of Boa Vista (Roraima) [47]. Despite quarantine efforts, this mite became established in South America inasmuch as in 2011; it was reported in the urban areas of Manaus occurring not only on coconut plants but also on dwarf royal palm (*Veitchia merrillii* (Becc.) H. E. Moore) and fishtail palm tree (*Caryota mitis* Lour) [55]. Recently, in May 2015, the RPM was found in the urban area of Dracena (state of São Paulo), about 2300 km southeast of Manaus, on several arecaceous plants such as *C. nucifera*, *Phoenix roebelenii* O'Brien, and *Rhapis excelsa* (Thunb.) A. Henry [54].

Discovery of *R. indica* in several countries in South America suggests that this region exhibits climate conditions, which, along with the wide diversity of host plant species, stimulate its development, representing an imminent threat to the economy of those countries where coconut palm and banana are grown as crops of economic importance.

4. Conclusions

Biological invasions have increased greatly in the last century due to the intensification in international trade, thus representing one of the most relevant threats for biodiversity in agroecosystems. Over several decades, scientists are more interested in phytophagous mites since some noneconomic species have become severe pests on many crops due to wrong pest management strategies. Thus, phytophagous mites from the Neotropical region such as the cassava green mite, the coconut mite, and the tomato spider mite have been introduced in the Old World. As expected, some mite species have also been introduced in the Neotropical region, i.e., *S. spinki*, *S. hindustanicus*, and *R. indica* with remarkable economic impact on agriculture. These biological invasions in the New World require the participation of several public institutions (Universities, Government Agricultural Institutions) and farmers in order to mitigate the current impact on production of rice, citrus, coconut, and Musaceous crops. Most of the research has been focused on geographical distribution, host plant range, and natural enemies associated, but few studies have dealt with management strategies.

Acknowledgements

The authors would like to acknowledge the Dirección de Investigación y Desarrollo (DIDE) of Universidad Técnica de Ambato and the Centre for Agricultural Bioscience International (CABI) for the financial support to this research.

Conflict of interest

The authors have declared no conflict of interests.

Author details

Carlos Vásquez[1*] and Yelitza Colmenárez[2]

1 Faculty of Agronomy Sciences, Technical University of Ambato, Ambato, Ecuador

2 CABI Brazil, UNESP- Fazenda Experimental Lageado, Fundação de Estudos e Pesquisas Agrícolas e Florestais, Botucatu-São Paulo, Brazil

*Address all correspondence to: ca.vasquez@uta.edu.ec

References

[1] Blackburn TM, Essl F, Evans T, Hulme PE, Jeschke JM, Kühn I, et al. A unified classification of alien species based on the magnitude of their environmental impacts. PLoS Biology. 2014;**12**(5). DOI: 10.1371/journal.pbio.1001850

[2] Pysek P, Jarosik V, Hulme P, Pergl J, Hejda M, Schaffner U, et al. A global assessment of invasive plant impacts on resident species, communities and ecosystems: The interaction of impact measures, invading species' traits and environment. Global Change Biology. 2012;**18**:1725-1737. DOI: 10.1111/j.1365-2486.2011.02636.x

[3] Ricciardi A, Hoopes MF, Marchetti MP, Lockwood JL. Progress toward understanding the ecological impacts of nonnative species. Ecological Monographs. 2013;**83**(3):263-282

[4] Colautti RI, MacIsaac HJ. A neutral terminology to define 'invasive' species. Diversity and Distributions. 2004;**10**:135-141

[5] Paini DR, Sheppard AW, Cook DC, De Barro PJ, Worner SP, Thomas MB. Global threat to agriculture from invasive species. Proceedings of the National Academy of Sciences of the United States of America. 2016;**113**(27):7575-7579. DOI: 10.1073/pnas.1602205113

[6] Preston CD, Pearman DA, Hall AR. Archaeophytes in Britain. Botanical Journal of the Linnean Society. 2004;**145**:257-294

[7] Ferragut F, Navia D, Ochoa R. New mite invasions in citrus in the early years of the 21st century. Experimental & Applied Acarology. 2012;**59**(1-2):145-164. DOI: 10.1007/s10493-012-9635-9

[8] Navia D, de Moraes GJ, Flechtmann CHW. Phytophagous mites as invasive alien species: Quarantine procedures. In: Proceedings of the XI International Congress of Acarology. Merida: Universidad Nacional Autonoma de Mexico; 2007. pp. 307-316

[9] Jeschke JM, Bacher S, Blackburn TM, Dick JT, Essl F, Evans T, et al. Defining the impact of non-native species. Conservation Biology. 2015;**28**(5):1188-1194. DOI: 10.1111/cobi.12299

[10] Courtois P, Figuieres C, Mulier C, Weill J. A cost-benefit approach for prioritizing invasive species. Ecological Economics. 2018;**146**:607-620. DOI: 10.1016/j.ecolecon.2017.11.037

[11] Martins CBC, Almeida LM, Zonta-de-Carvalho RC, Castro CF, Pereira RA. *Harmonia axyridis*: A threat to Brazilian Coccinellidae? Revista Brasileira de Entomologia. 2009;**53**(4):663-671. DOI: 10.1590/S0085-56262009000400018

[12] Richardson DM, Pysek P, Rejmánek M, Barbour MG, Panetta FD, West CJ. Naturalization and invasion of alien plants: Concepts and definitions. Diversity and Distributions. 2000;**6**:93-107

[13] Beck KG, Zimmerman K, Schardt JD, Stone J, Lukens RR, Reichard S, et al. Invasive species defined in a policy context: Recommendations from the federal invasive species advisory committee. Invasive Plant Science and Management. 2008;**1**:414-421. DOI: 10.1614/IPSM-08-089.1

[14] Bolland H, Gutiérrez J, Fletchmann C. World Catalogue of the Spider Mites Family (Acari: Tetranychidae). Leiden: Koninklijke Brill NV; 1998. p. 392

[15] Nienstaedt B, Marcano R. Estudio de la biología del ácaro hindú de los cítricos *Schizotetranychus hindustanicus* (Hirst, 1924) (Acari: Tetranychidae), en tres tipos de alimentos. Entomotropica. 2009;**24**(2):51-56

[16] Quirós M, Geraud-Pouey F. *Schizotetranychus hindustanicus* (Hirst) (Acari: Tetranychidae), new spider mite pest damaging citrus in Venezuela, South America. In: Proceedings of the XI International Congress of Acarology; 8-13 September 2002. Mérida: Universidad Nacional Autónoma de México; 2008. pp. 255-256

[17] Návia D, Marsaro AL Jr. First report of the citrus Hindu Mite, *Schizotetranychus hindustanicus* (Hirst) (Prostigmata: Tetranychidae), in Brazil. Neotropical Entomology. 2010;**39**(1):140-143

[18] Quirós M, Geraud-Pouey F. *Schizotetranychus hindustanicus* (Hirst) (Acari: Tetranychidae), new spider mite pest damaging citrus in Venezuela, South America. In: International Congress of Acarology. Mérida, México; 8-13 September 2002; Program and Abstract Book; 2002. pp. 255-256

[19] Nienstaedt BM. Estudio de algunos aspectos biológicos y ecológicos del ácaro hindú de los cítricos *Schizotetranychus hindustanicus* (Hirst, 1924) (Acari: Tetranychidae) en Maracay, Venezuela [thesis]. Maracay: Universidad Central de Venezuela; 2007

[20] Quirós M, Dorado I. Eficiencia de tres productos comerciales en el control del ácaro hindú de las cítricas *Schizotetranychus hindustanicus* (Hirst), en el laboratorio. In: Congreso Venezolano de Entomología; 4-7 July 2005; Maracaibo. Maracay: Entomotropica; 2005. pp. 184-185

[21] Mesa NC. Ácaros Asociados a Cítricos en Colombia [Internet]. 2010. Available from: http://www.comfenalcoantioquia.com/boletinempresas/MEMORIAS_CITRICOS/CONFERENCISTAS/NACIONALES/Nohora%20Cristina%20Mesa%20-%20Colombia/Acaros%20en%20citricos%20-%20Congreso.pdf

[22] Nienstaedt B, Marcano R. Fluctuación poblacional y distribución vertical del ácaro *Schizotetranychus hindustanicus* (Hirst, 1924), sobre especies de *Citrus*. Entomotropica. 2009b;**24**(2):57-63

[23] Alvarado G, Fréitez F. *Schizotetranychus paezi* sp. n. y *Schizotetranychus oryzae* (Acarina: Tetranychidae) atacando arroz en Venezuela. Agronomía Tropical. 1976;**26**(2):159-165

[24] Zhang Z, Zhang Y, Lin J. Mites of *Schizotetranychus* (Acari: Tetranychidae) from moso bamboo in Funjian, China. Systematic and Applied Acarology Special Publications. 2000;**4**:19-35. DOI: 10.11158/sasp.4.1.5

[25] Almaguel L, Cabrera I, Hernández J, Ramos M, Sandoval I. Etiología, Biología, Ecología y Manejo Integrado del Vaneado de la Panícula y Pudrición de la Vaina del Arroz en Cuba. Instituto de Sanidad Vegetal, Instituto de Investigaciones del Arroz, Centro Nacional de Sanidad Agropecuaria: La Habana; 2002

[26] Chandrasena GDSN, Jayawardane JDKM, Umange SD, Gunawardana ADBU. Host range of panicle rice mite *Steneotarsonemus spinki* Smiley (Acari: Tarsonemidae) in Sri Lanka. Universal Journal of Agricultural Research. 2016;**4**(1):21-24. DOI: 10.13189/ujar.2016.040104

[27] Rao J, Prakash A. Paddy field weed, *Schoenoplectus articulatus* (Linn.) Palla (Cyperaceae): A new host of tarsonemid mite, *Steneotarsonemus spinki* Smiley and panicle thrips, *Haplothrips ganglbaureri* Schmutz. Journal of Applied Zoological Researches. 2002;**13**:174-175

[28] Hummel NA, Castro BA, McDonald EM, Pellerano MA, Ochoa R. The panicle rice mite, *Steneotarsonemus spinki* Smiley, a re-discovered pest of rice in the United States. Crop

Protection. 2009:1-14. DOI: 10.1016/j.cropro.2009.03.011

[29] Shikata E, Kawano S, Senboku T, Tiongco ER, Miyajima K. Small virus-like particles isolated from the leaf sheath tissues of rice plants and from the rice tarsonemid mites, *Steneotarsonemus spinki* Smiley (Acarina, Tarsonemidae). Annals of the Phytopathological Society of Japan. 1984;**50**:368-374

[30] Ramos M, Rodríguez H. *Steneotarsonemus spinki* Smiley (Acari: Tarsonemidae): Nuevo informe para Cuba. Revista de Protección Vegetal. 1997;**13**(1):25-28

[31] Ramos M, Gómez C, Cabrera RI. Presencia de *Steneotarsonemus spinki* (Acari: Tarsonemidae) en cuatro variedades de arroz en la República Dominicana. Revista de Protección Vegetal. 2001;**16**:6-9

[32] Sanabria C, Aguilar H. El Ácaro del Vaneo del Arroz (*Steneotarsonemus spinki* L: Tarsonemidae). Ministerio de Agricultura y Ganaderia: San José de Costa Rica; 2005. 16 p

[33] Herrera LAR. Ácaro del Vaneamiento del Arroz-*Steneotarsonemus spinki* Smiley (Prostigmata: Tarsonemidae) [Internet]. 2005. Available from: www.flar.org/pdf/foro-agosto-pdf05/acaro.pdf

[34] Almaguel L, Botta E. Manejo Integrado de *Steneotarsonemus spinki* Smiley: Resultados de Cuba y Transferencia Para la Región de Latinoamérica y el Caribe [Internet]. 2005. Available from: http://www.inisav.cu/OtrasPub/Curso%20acarolog%C3%ADa.pdf

[35] Castro BA, Ochoa R, Cuevas FE. The threat of the panicle rice mite, *Steneotarsonemus spinki* Smiley, to rice production in the United States. In: Proceedings of the Thirty First Rice Technical Working Group; 26 February-1 March 2006; Texas: The Woodlands. pp. 97-98

[36] Arriaga JT. Detection of the Rice Tarsonemid Mite (*Steneotarsonemus spinki* Smiley) in Palizada, Campeche, Mexico [Internet]. 2007. Available from: http://www.pestalert.org/oprDetail.cfm%3FoprID¼4268

[37] ICA (Instituto Colombiano Agropecuario). Resolución No. 001195 de 2005. Diario Oficial, Edición 45.892. 27 abril de 2005 [Internet]. 2005. Available from: https://normograma.info/invima/docs/pdf/resolucion_ica_1195_2005.pdf

[38] Aguilar H, Murillo P. Nuevos hospederos y registros de ácaros fitófagos para Costa Rica: Periodo 2002-2008. Agronomica Costarricense. 2008;**32**(2):7-28

[39] Mendonça RS, Návia D, Flechtmann CHW. *Raoiella indica* Hirst (Acari: Prostigamata: Tenuipalpidae), o ácaro vermelho das palmeiras: Uma ameaça para as Américas. Documento 146. Brasilia: EMBRAPA; 2005. 37 p. DOI: 10.13140/RG.2.1.1194.0886

[40] Flechtmann CHW, Etienne J. The red palm mite, *Raoiella indica* Hirst, a threat to palms in the Americas (Acari: Prostigmata: Tenuipalpidae). Systematic and Applied Acarology. 2004;**9**:109-110. DOI: 10.11158/saa.9.1.16

[41] Kane E, Ochoa R, Mathurin G, Erbe E. *Raoiella indica* Hirst (Acari: Tenuipalpidae): An island-hopping mite pest in the Caribbean [Internet]. 2005. Available from: www.sel.barc.usda.gov/acari/PDF/Raoiella%20indica-Kane%20et%20al.pdf

[42] Etienne J, Flechtmann CHW. First record of *Raoiella indica* (Hirst, 1924) (Acari: Tenuipalpidae) in Guadeloupe and Saint Martin, West Indies. International Journal of

Acarology. 2006;**32**:331-332. DOI: 10.1080/01647950608684476

[43] Rodrigues JCV, Ochoa R, Kane E. First report of *Raoiella indica* Hirst (Acari: Tenuipalpidae) and its damage to coconut palms in Puerto Rico and Culebra Island. International Journal of Acarology. 2007;**33**:3-5. DOI: 10.1080/01647950708684493

[44] Welbourn C. Red Palm Mite *Raoiella indica* Hirst (Acari: Tenuipalpidae) [Internet]. 2007. Available from: http://www.doacs.state.fl.us/pi/enpp/ento/r.indica.html

[45] Vásquez C, Quirós M, Aponte O, Sandoval MF. First report of *Raoiella indica* Hirst (Acari: Tenuipalpidae) in South America. Neotropical Entomology. 2008;**37**(6):739-740. DOI: 10.1590/S1519-566X2008000600019

[46] Carrillo D, Návia D, Ferragut F, Peña JE. First report of *Raoiella indica* (Acari: Tenuipalpidae) in Colombia. Florida Entomologist. 2011;**94**(2):370-371. DOI: 10.1653/024.094.0241

[47] Návia D, Marsaro AL Jr, da Silva FR, Gondim MGC Jr, de Moraes GJ. First report of the red palm mite, *Raoiella indica* Hirst (Acari: Tenuipalpidae), in Brazil. Neotropical Entomology. 2011;**40**(3):409-411. DOI: 10.1590/S1519-566X2011000300018

[48] Flechtmann CHW, Etienne J. Un nouvel acarien ravageur des palmiers en Martinique, premier signalement de *Raoiella indica* pour les Caraïbes. Phytoma. 2005;**548**:10-11

[49] Beard JJ, Ochoa R, Bauchan GR, Welbourn WC, Pooley C, Dowling APG. External mouthpart morphology in the Tenuipalpidae (Tetranychoidea): *Raoiella* a case study. Experimental & Applied Acarology. 2012;**57**:227-255. DOI: 10.1007/s10493-012-9540-2

[50] Vásquez C, Egurrola Z, Valera R, Sanabria ME, Colmenárez Y. Anatomía y química foliar en especies ornamentales de Arecaceae: Posibles barreras a la alimentación de *Raoiella indica* Hirst (Acari: Tenuipalpidae). Gayana Botánica. 2015;**72**(2):256-264. DOI: 10.4067/S0717-66432015000200012

[51] Roda A, Dowling A, Welbourn C, Peña J, Rodrigues JC, Hoy M, et al. Red palm mite situation in the Caribbean and Florida. Proceedings of the Caribbean Food Crops Society. 2008;**44**:80-87

[52] Vásquez C, de Moraes GJ. Geographic distribution and host plants of Raoiella indica and associated mite species in northern Venezuela. Experimental & Applied Acarology. 2013;**60**:73-82. DOI: 10.1007/s10493-012-9623-0

[53] Carrillo D, Amalin D, Hosein F, Roda A, Duncan RE, Peña JE. Host plant range of *Raoiella indica* (Acari: Tenuipalpidae) in areas of invasion of the new world. Experimental & Applied Acarology. 2012;**57**:271-289. DOI: 10.1007/s10493-011-9487-8

[54] Oliveira DC, Prado EP, de Moraes GJ, de Morais EGF, Chagas EA, Gondim CMG Jr, et al. First report of *Raoiella indica* (Acari: Tenuipalpidae) in southeastern Brazil. Florida Entomologist. 2016;**99**(1):123-125. DOI: 10.1653/024.099.0124

[55] Rodrigues JCV, Antony LMK. First report of *Raoiella indica* (Acari: Tenuipalpidae) in Amazonas state, Brazil. Florida Entomologist. 2011;**94**(4):1073-1074. DOI: 10.1653/024.094.0452

Chapter 8

Biology and Ecology of Some Predaceous and Herbivorous Mites Important from the Agricultural Perception

Muhammad Sarwar

Abstract

Mites are numerous species of minute arthropods, members of class Arachnida subclass Acari or Acarina and pests of many economic prominence living in a wide range of habitats. Mites are predators and parasites, performing crucial means of biological control, essential herbivores and detritivores, acting fungivorous and saprophytic, vectors of diseases, and play vital role in soil formation. These live on plants and animals, in the depths of ocean, in soil and fresh or brackish water, in lungs of birds and animals, in stored grains and stored products, on leaves of rainforest, and in human clothes and bedding. In spite of magnificent diversity of predaceous, phytophagous and granary mites found on plants and stored grains, these are often overlooked, and even skilled zoologists may be unaware of their importance. This chapter aims to provide an updated analysis of their biology, life history, reproduction and ecology to fill gap in our understanding of these fascinating creatures for pests controlling.

Keywords: arthropod, Acari, mite, pest, Phytoseiidae, control

1. Introduction

Mite complex is worldwide in its distribution in all regions of globe and more prominent in tropical a well as subtropical climates. Mites can be either inflicting damage to humans and animals [1, 2], or pestilent that feed on plants [3] and stored commodities [4–23], otherwise predacious which are the carnivorous of leaf-feeding mites and other pests [24]. All harmful types of mites are able to devastate agricultural crops, fruits and vegetables [25, 26]. During the previous few decades, owing to increasing concerns over health, environment and pest resistance risks accompanying with chemical control, and the use of alternate pest management strategies has received considerable attention [27, 28]. In this context, the uses of generalist predators that can perform as a broad spectrum fighters against pests have been greatly encouraged [29–31].

Currently, mites belonging to the family Phytoseiidae (Arachnida: Mesostigmata) are economically important predators of some phytophagous mites and insects in greenhouses or field crops. Amongst others predators, mass reared phytoseiid mites are commercially available and used, against spider mites, thrips and whiteflies infestations on plants. Phytoseiid mites use odors (kairomones) associated with mite-infested plants to locate their prey or when predators contact spider mite webbing, these intensify their search for prey and may identify prey eggs and distinguish these from non-prey objects. The existence of a water-soluble feeding stimulant on prey eggs as well is postulated [32].

Predator mites which fit to the family Phytoseiidae, are categorized by long legs, with the front pair pointing frontward and comparatively have few hairs (<20 pairs) on their back. The color of mites can differ from deep red to pale yellow liable to the prey items eaten. Mites that feed on whiteflies and thrips are commonly pale yellow to pale tan. Phytoseiid mites have five life stages in life cycle like egg, larva, protonymph, deutonymph and adult. Most mites of this family are free-living predators in the deutonymphal and adult stages on a variety of arthropods in plants or crops. This chapter presents broad-spectrum ideas of the findings in research focusing on rhetorical aspects of biology and ecology of some predacious as well as harmful mites with particular reference to their possible role in biological control [33, 34].

2. Beneficial mites

Beneficial mites are excellent biological control agents and have been used in controlling of tiny mite pests and insect pests that cause a serious damage to many economically important crops.

2.1 *Neoseiulus* (*Amblyseius*) *cucumeris* (Oudemans)

Cucumeris predatory mite *Neoseiulus* (= *Amblyseius*) *cucumeris* (Oudemans) has a worldwide distribution because of its natural occurrence and commercial use in various parts of the ecosphere. Cucumeris is soft-bodied and translucent pale brown to sometimes tan-colored depending on the food consumed. These can be distinguished from most pest mites by their shape and mobility (**Figure 1**). Cucumeris moves rapidly along the underside of leaves and in flowers of plants. Adult mite is pear-shaped and may range between 0.5 and 1.0 mm in length with long legs. Gnathosoma corniculi is slightly convergent distally, dorsal shield strongly reticulate with 17 pairs of setae, and sternal shield smooth with few lateral striae, three pairs of setae and two pairs of lyrifissures. Eggs are laid on the leaf surface, on domatia or on hairs (trichomes) along the midrib on the underside of leaves and occasionally on petiole hairs. The eggs are round, translucent white and measure 0.14 mm in diameter. Females lay 1–3 eggs per day for an average of 35 eggs over their lifetime. The eggs hatch in about 3 days later [35].

Figure 1.
Neoseiulus cucumeris.

Cucumeris populations have somewhat more females than males (64% females). Mite develops through one larval stage and two nymphal stages (protonymph and deutonymph) before becoming adults. The non-feeding larvae emerge from eggs in about 3 days and molt into protonymphs 2 days later. The two nymphal stages last for 7–10 days before developing into adults. Adults live for up to 30 days and eat an average of 1 thrips/day. Cucumeris has a life cycle of 10–12 days at 20°C, while development time at 75°F is 6–9 days and development takes from 8 to 11 days (at 20–25°C). Cucumeris prefers environment with >65% relative humidity (R. H.), but eggs can survive at as low as 40% R. H. Greenhouse *N. cucumeris* is not susceptible to diapause, however, less effective above 28°C. Optimum temperature range for Cucumeris mites is 66–80°F with humidity between 65 and 75%. The lowest developmental time of the predator from neonate larva to adult emergence (7.50 days), the highest immature survivorship (99%), higher fecundity (3.85 eggs/female/day), long oviposition period (26.57 days), and the highest adult female's life span (39 days) have been observed when offered maize pollens along with *Tyrophagus putrescentiae* (Schrank) as prey [36].

For studying food habits of predatory mite *N. cucumeris*, well known pests, stored food mite *T. putrescentiae*, red spider mite *Tetranychus urticae* Koch and western flower thrips *Frankliniella occidentalis* (Pergande), have been used as prey. Significant differences have been observed for the types of prey diet used, *T. putrescentiae* proved the most suitable prey closely followed by *T. urticae* and then *F. occidentalis* diet, wherein, duration of the developmental stages of the predatory mites noted 7.6, 7.7 and 8.5 days, respectively [37].

Cucumeris is an aggressive predator of several soft-bodied pests and generally microclimates inside the greenhouse crop seem to be significant for their existence. Cucumeris feeds on little (first and second instar) thrips on foliage and flowers, and does not nourish on big larvae or adult thrips. The prime targets of Cucumeris are thrips species including western flower thrips (*F. occidentalis*), onion thrips (*Thrips tabaci* Lind.), plague thrips (*Thrips imaginis* Bagnall), melon thrips (*Thrips palmi* Karny), common blossom thrips (*Frankliniella schultzei* Trybom) and chilli thrips (*Scirtothrips dorsalis* Hood) [38]. These are known to feed on the immature stages of a variety of plant damaging mites, for instance, broad mite (*Polyphagotarsonemus latus* Banks), cyclamen or strawberry mite (*Phytonemus pallidus* Banks), spider mites, tomato russet mite and mites of the genus *Schizotetranychus* in ornamental, fruit and market garden crops. Cucumeris is also used effectively for control of thrips in stored red salad onions [39, 40].

Cucumeris is an appropriate enemy for many tiny pests of greenhouse crops, and both outdoor and indoor strawberry crop. This is able to live on pollen in the absence of pest and as a result might be used precautionary in crops such as strawberries or capsicums that produce pollen. Cucumeris has been efficaciously used for thrips control in capsicums, cucumbers, berry fruits and eggplants as well as in ornamental crops such as rose and gerbera, and other potted plants. In circumstances having very huge thrips pressure, Cucumeris ought to be always used in combination with the predatory pirate bug *Orius tantillus* (Motschulsky). Cucumeris is not efficient control agent on geraniums or tomatoes owing to leaf structure and toxic plant exudates [41].

In recent years, various delivery systems (formulations) of *N. cucumeris*, such as buckets (100,000 mites) or bottle (50,000 mites) containers are offered for direct release in greenhouse, field and nursery operations. Both systems comprise of predatory mites and bran mites *T. putrescentiae* (a temporary food source of the predatory mite) mixed with bran or vermiculite. Cucumeris is supplied from the insectary at approximately 100,000 predators per liter of substrate. Each liter of insectary material should cover 600–2500 m^2 of greenhouse area (4–16 L/ha). For strawberries, release is done at the rate of 2 L/10,000 plants at initial flowering in spring and yet again 4–6 weeks later. Wherever thrips are high, a third release of predator can be desirable. In greenhouse crops, Cucumeris must be released at a rate of 50–200 predators/m^2 of cropping area (4–16 L/ha) subjected to thrips density. In crops with little thrips density and which have enough pollen, for instance, capsicums (once flowering has started), the lesser amount can be used, however must be used in two doses about 14 days apart to make certain even dispersal through the crop. Sprinkle 200 mL of mixture per ton of onions over the top of the onions once stored into bins prior to storage [42].

2.2 *Neoseiulus* (= *Amblyseius*) *fallacis* (Garman)

Mite *Neoseiulus* (= *Amblyseius*) *fallacis* (Garman) is an excellent general predator for control of many different types of mites in warm and moderately humid environments. Particularly, it targets two-spotted spider mite (*T. urticae*), broad mite, hemp russet mite, European red mite (*Panonychus ulmi* Koch), spruce spider mite (*Oligonychus ununguis* Jacobi), southern red mite (*Oligonychus ilicis* McGregor), bamboo mite (*Schizotetranychus celarius* Banks), cyclamen mite, other small arthropods and is also able to feed on an array of pests, especially their eggs on roses and vegetable crops. Fallacis can also be effective in orchards as it tolerates higher temperatures and lower humidities. For the reason that *N. fallacis* is a hungry user of mites and owing to its density rises rapidly relatively to its prey, it is able to surpass an intensifying pest populations. Additionally, it progresses into the mature stage in nearly one third the time needed by other mite predators. Moreover, this may also feed on pollen and can survive for periods having pollen alone, which makes it an excellent preventative natural enemy [43].

Adults of *N. fallacis* are about 0.5 mm long, with pear-shaped bodies. These can vary in color from cream to orange-beige, shiny and semi-transparent, with long legs (**Figure 2**). The immature stages are generally a semi-transparent cream color and difficult to see without a microscope. The eggs are oval, almost transparent, and 0.33 mm in diameter (double than size of a two-spotted spider mite egg). Adult females lay 1–5 eggs per day along the ribs on the undersides of leaves and a total of 26–60 eggs over their lifetime (which could be between 14 and 62 days) are laid. The eggs hatch in 2–3 days and newly hatched predators do not eat, but later stages and adults feed on all stages of prey. Generally, female *A. fallacis* can eat 2–16 spider mites per day. Of the five *N. fallacis* life stages, only the first nymphal stage is six legged, while all other post-egg stages have eight legs [44, 45].

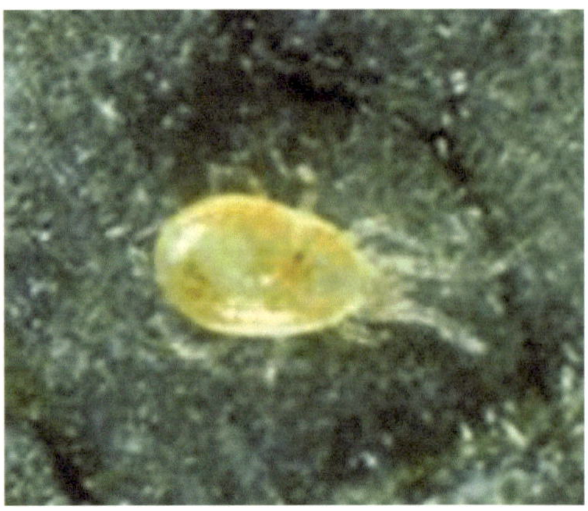

Figure 2.
Neoseiulus fallacis.

Growth from egg to adult takes place in 7–9 days at 70°F, 3 days at 85°F and at 78°F a fourfold rise in numbers can occur within 4 days. Under optimum conditions in the field, densities may increase from 10 predators per 100 leaves to 200–500 predators per 100 leaves in just 2 weeks. Adult mated females enter diapause in response to the short days in the fall (<14 hours of daylight) in plant crevices or other protected areas. As a result, these stop reproducing and move into sheltered areas, such as under bark or ground cover. But, these do not enter diapause in greenhouses or interior plantscapes if the temperature is 64°F (18°C) or above. These emerge as early as bloom, but in reduced numbers due to heavy winter mortality. Fallacis increases in number rapidly and adults become numerous by July or August, and on an average 40–60 eggs are laid. Warmer or cooler conditions accelerate or slow down reproduction/feeding, respectively, and these live about 20 days [46, 47].

Mite predator *N. fallacis* works extremely well to control mite infestations in greenhouses, so, it is suggested for use in tomatoes, roses and other vegetable crops. On field crops, before introducing *N. fallacis*, monitoring and counting should be done to determine numbers of spider mites and existing predators. Inoculate only those fields having spider mite populations of 0.3 mites per leaf and higher [48]. Spread *N. fallacis* evenly throughout the field using 150–200 release points per hectare (60–80 per acre), concentrating extra predators near to higher mite counts. For new strawberries and mint or raspberries and currants plantings, release 25,000 predators per hectare (10,000 per acre) as soon as possible after planting or 10 days after applying insecticides to control aphids, and on producing fields, release 17,000 per hectare (7000 per acre).

2.3 *Phytoseiulus persimilis* Athias-Henriot

Predator mite *Phytoseiulus persimilis* Athias-Henriot, is a specific predator of web-spinning spider mites like two spotted spider mite. Indeed, *P. persimilis* nourishes, breeds, and completes growth merely on mites in the subfamily Tetranychinae, even though it too feeds on young thrips and may be cannibalistic at what time spider mite prey is absent [49]. This species is approximately 0.5 mm long, fast moving, orange to bright reddish orange in color, has a teardrop-shaped body, long legs and is slightly larger than its prey (**Figure 3**). Adult females are reddish, pear-shaped and active at room temperature. Immatures and males are smaller and lighter in color. The life cycle of *P. persimilis* has been determined under a diurnal temperature cycle of 58–83°F. Eggs are oval, oblong and approximately twice as large as the pest mite eggs and hatch in 2–3 days. The adult female may lay up to 60 eggs during its 50 day-long lifetime at 17–27°C. The adult female, after a pre-ovipositional (the time between emergence from the egg to the deposition of the first egg) period of 3.0 days, laid an average of 2–4 eggs per day for 22.3 days. The average duration of incubation for both males and females is 3.1 ± 0.2 days [50].

Figure 3.
Phytoseiulus persimilis.

Larvae do not attempt to feed and remain inactive near the old egg shell. Although the larval stage does not feed, yet the subsequent nymphs and adults feed on all stages of prey. Both males and females remained in the larval stage for an average of 1.0 ± 0.1 days. Immatures are normally pale salmon in color. The male and female protonymphal stages lasted 1.7 and 1.6 days, respectively. During this time both males and females ate an average of 4–4 eggs of *T. urticae*. The eight-legged final nymphal stage (deutonymphal stage) remained active throughout the period. Both males and females remained an average of 1.7 ± 0.1 days in this stage, and during that time the deutonymphs ate 6.0 eggs. Both males and females start to forage just about instantaneously afterward these have molted. Afterwards nourishing, the females become very active and spend plentiful of its time running around the cell enclosure. Simply, when a male is faced and mating has taken place then the female settles down. Usually 6–12 hours is elapsed between the time of molting and the time of mating. The duration of mating is erratic, taking from 5 minutes to several hours. The duration of the ovipositional period (period during which eggs are deposited) ranged from 6 to 39 days, with an average of 22.3 days. The maximum number of eggs laid by any female in a day is six. During the pre-ovipositional and ovipositional periods, the females consumed 7.3 and 14.3 eggs of *T. urticae*, respectively. After ceasing to oviposit, the females lived an average of 7.1 days during which each individual consumed 3.9 eggs per day. Each sex ate an average of 10.5 eggs during entire development. Total developmental time for males is 7.5 days and for females 7 days. However, total generation time from egg to adult ranges from 25.2 days at 15°C (59°F) to 5.0 days at 30°C (86°F) [51–53].

Due to its tropical origin, *P. persimilis* does not have a diapause stage and is active year-round in enclosed habitats such as interior plantscapes and greenhouses. Because this mite is much efficient hunter and disperser, it can cause effective extinction of its spider mite prey. Persimilis predaceous mite, is one of the pillars of greenhouse integrated pest management programs for control of spider mites on vegetables and ornamentals [54, 55].

2.4 *Neoseiulus* (= *Amblyseius*) *californicus* (McGregor)

Predaceous mite *Neoseiulus californicus* (McGregor) is a predator, which primarily attacks spider mites, but also feeds more generally on many leaf inhabiting mites or other small insects and pollen. It is tolerant and active at both high and low temperatures, and low humidity, but prefers warm to hot conditions where it can reproduce very quickly.

Mite *N. californicus* is pear-shaped, buff to tan in color and <1 mm long. It is nearly related in look to *N. cucumeris*. Males are greatly smaller and darker brown than females, with females commonly more abundant (**Figure 4**). The female lays ovate, pure white eggs singly or in clusters on underside of leaf, frequently on leaf hairs or else at the junction of veins. The laid eggs are bigger than those of spider mites. These hatch 1–2 days afterward and pass through a 6-legged larval and two 8-legged nymphal stages to become mature. The life-cycle from egg to adult is temperature reliant, however under greenhouse situations it varies from 4 to 10 days. The adult females live about 20 days and lay three eggs per day, commencing after 2–3 days. Pollen is its maintenance food only, however this can helpful for establishment and persistence in the crop at what time pests are short. In the deficiency of food, predator scatters over the crop and on the ground examining for new prey, however a lot of will halt in crop and wait for the appearance of new prey [56, 57].

Predator *N. californicus* is a promising agent for successful *T. urticae* (two-spotted spider mite, red spider mite) control through conservation techniques in strawberry; *Panonychus citri* McGregor (citrus red mite); *Tarsonemus pallidus* Banks

Figure 4.
Neoseiulus californicus.

(cyclamen mite, strawberry mite); *P. latus* (broad mite, yellow tea mite, citrus silver mite); *Raoiella indica* Hirst (red palm mite); *Brevipalpus* spp.; *Panonychus ulmi* (European red mite, fruit tree red spider mite); *Tetranychus cinnabarinus* (Boisduval) (carmine spider mite, cotton red spider mite) in flowering crops; fruit-bearing vegetables; ornamentals (protected); spice crops; berries; and grapes. Promoting preservation of *N. californicus* can furthermore be done on an extensive range of crops. Appropriate crops comprise ornamentals (gerbera, chrysanthemum, rose), vegetables (capsicum, eggplant, cucumber) and herbs. It is frequently used in greenhouse crops production, however, can as well be used in the field, mostly fruit crops like as pome, melons and stone fruits [58].

Californicus works in the superlative form while used preventatively, or else when spider mites are initially observed in the crop. It establishes the best early in the crop and when is permitted to build up prior to spider mites found. Predator *P. persimilis* ought to be used for quicker knockdown of spider mites for the period of modest temperature and high humidity circumstances, however wherever situations are identified to be hot or dry later in the year, it is recommended to discharge and settle the species early in the crop cycle. If spider mite numbers are already high, it is desirable the usage of a non-disruptive miticide to lower the population prior to release or after establishment for helping in pest control. Residual broad spectrum insecticides should not be used for at least 4 weeks prior to predator release. Use *N. cucumeris* preferentially for broad and cyclamen mites, though *N. californicus* will assist in pest control and may be more effective at higher temperatures [59, 60].

Californicus are primarily sent in a loose, vermiculite-based medium and the predator should be distributed evenly through the crop on foliage, with additional material at ends of rows and in hotter areas prone to spider mites. Rates will vary depending on the crop and infestation level, however, the subsequent rates have been determined as preventative @ 25/m^2 (2.5 L/ha) releasing 2 weeks apart for 2–3 releases, and after spider mite detection 100–200/m^2 (10 L/ha) weekly for at least three applications.

2.5 Swirski mite *Amblyseius swirskii* (Athias-Henriot)

The species *Amblyseius* (*Typhlodromips*) *swirskii* (Athias-Henriot) is considered a generalist predator and readily consumes small soft-bodied arthropod pest species as well as pollen or plant exudates. It can be found on various crops including apples, apricot, citrus, vegetables and cotton. It has attracted substantial interest as a biological control agent of mites, thrips and whiteflies in greenhouses and nursery crops. This species is documented to feed and reproduce on a wide range of prey from several orders, including thrips (western flower, onion, melon and chili), whiteflies (greenhouse and silverleaf) [61], and plant feeding mites (spider, broad and eriophyoid) [62, 63]. It mainly feeds on the immature stages of thrips and whitefly, although it also attacks the adult stages of smaller pest species. A recent study documented *A. swirskii* feeding and probably reproducing on immature Asian citrus psyllids. Adult predatory mites search for their prey or wait for it to pass by and then suck it dry [64].

Adults are pear-shaped, 0.5 mm in length with an unsegmented body and four pairs of long legs, and males may be slightly smaller than females (**Figure 5**). The eggs are round and transparent white and measure approximately 0.15 mm in diameter. These mites lay their eggs on leaf hairs (trichomes) and along the veins on the inner surface of leaves mainly at the intersection of main and lateral ribs. Females prefer to lay eggs on leaf hairs on the underside of plant leaves near plant domatia (small hairy tufts or pockets found on the lower surface of some leaves), which may be an adaptation to avoid egg from predators. The eggs hatch in about 3 days later. Larvae are pale white to nearly transparent in color and only have three pairs of legs. Mobile stages are beige-pink, droplet shaped and 'pushed down' position on short legs. The protonymph (second stage) and deutonymph (third stage) have four pairs of legs and are darker than the larvae. All stages can be found in the corner of main vein and lateral veins, and in the flowers [65, 66].

In addition to arthropod prey, *A. swirskii* can survive and reproduce on various pollens and gain nutrition from plant nectars, which may allow them to persist during periods of low pest density and improve their effectiveness as biological control agents. Development of *A. swirskii* is influenced by type of food (prey, pollen and plant exudates) as well as availability of food and environmental conditions. Mites develop between 18 and 36°C and at 60% relative humidity [67]. By feeding on prey, the egg to adult development period at 25°C is around 5 days [68]. Mite *A. swirskii* feeding on live prey develops faster and lays more eggs when compared with pollen-feeding *A. swirskii*, for example, females laid 26 and 38 eggs on pollen and mite diets, respectively. When not actively hunting, *A. swirskii* is typically found on the underside of leaves along the midrib or in other protected locations such as domatia [69].

Species *A. swirskii* is commonly used to control whitefly and thrips in greenhouse vegetables (especially cucumber, pepper and eggplant) and some ornamental crops, and other pests on citrus and other subtropical crops. The mites are released directly in the crops in bran or vermiculite carriers sprinkled on the leaves or substrates, or may be broadcasted via air blast or other automated distribution technique. The recommended release rates are typically between 25 and 100 mites per m^2 depending on pest species, pest density and type of crop [70, 71].

Figure 5.
Amblyseius swirskii.

2.6 *Amblyseius* (*Neoseiulus*) *barkeri* (Hughes)

The predatory mite *Amblyseius barkeri* (Hughes) [= *A. mckenziei* Schuster and Pritchard] has played important roles in biological controls of thrips in glasshouses with cucumber and cabbage, and also feeds on tetranychid, tarsonemid and eriophyid mites, and pollen of various plants. In appearance, adults species of predaceous mites *A. barkeri* are small and reddish-brown in color (**Figure 6**). Mobile stages are pale-pink, droplet shaped and have 'pushed down position' on short legs. Mite *A. barkeri* in the field is indistinguishable from *N. cucumeris* and *N. californicus*. It is smaller than *N. cucumeris* and is therefore better suited as a predator for broad mites, which often live in tight and small crevices [72].

The eggs are oval, transparent white and around 0.15 mm in diameter. All stages can be mainly found in the corner of main vein and lateral veins, and in the flowers. When fed on *T. tabaci*, at 25°C, *A. barkeri* exhibited a mean time of 2.2, 0.8 and 3.2 days for the egg, larval and nymphal phases, with mortalities at 1.0, 1.0 and 3.1%, correspondingly. Females denoted 63% of the total population and needed several copulations for optimum fecundity. The oviposition time is 20.3 days and the mean oviposition rate 2.3 eggs per day, while the intrinsic rate of increase is 0.22 per day. The estimated life span is 29.6 days for male and 27.4 days for female. Both male and female *A. barkeri* devoured 3.3 nymphs of thrips per day (average value for the feeding stages), whereas the larva does not consume diet. Without thrips diet, *A. barkeri* is capable of consuming adults broad mite *P. latus*, two-spotted spider mite *T. urticae* and their eggs, and pollen of many plants. Cannibalism has been detected within *A. barkeri* at what time food is deficient [73, 74].

The predacious mite *A. barkeri* as well accomplished its life cycle when fed on eriophyid mite *Eriophyes dioscoridis* Soliman and Abou-Awad and tetranychid mite *T. urticae*. The growth is not accomplished while the phytoseiid mite fed on pollen grains of *Phoenix dactylifera* (L.), and *Ricinus communis* (L.), as an alternative food material. The adult female everyday devoured 17 and 102 individuals of *T. urticae* and *E. dioscoridis*, respectively. The day-to-day reproduction rate is 1.3 eggs per day after nourished on eriophyid mites, whereas the number of eggs laid improved (1.9 eggs per day) while fed on tetranychid mites [75].

It is used against various thrips species and broad mite as well as other tarsonemid mites in greenhouses and indoor plantscapes. It works well on *P. pallidus* (cyclamen mite) and *P. latus* mites (broad mite, yellow tea mite and citrus silver mite). Mites pierce the instars of their prey and suck out the internal organs to eat, but do not eat adults. It is found as wild populations on cut flowers (orchids,

Figure 6.
Amblyseius barkeri.

bromeliads and roses), on vegetable crops (capsicum) and strawberry. Adult predatory mites and nymphs suck young thrips larvae and multiple stages of broad mite. Predator can also eat pollen of peppers and eggplant and thus build its population in these crops [34].

2.7 *Amblyseius andersoni* (Chant)

Mite species *Amblyseius* (*Typhlodromips*) *andersoni* (Chant), is a polyphage, and the predator is able to consume alternative food sources such as young larvae of thrips, flower pollen, fungi and the sugary excretions of other pests. This predatory mite is ideal for vegetables, hardy ornamentals and fruit crops, for controlling red or two-spotted spider mite, fruit-tree red spider mite, and russet mites. An adult predatory mite has light beige color and a size of 0.5–1 mm long (**Figure** 7). The female of the mite lays several eggs every day on the underside of a leaf near its veins. Adult female mites lay eggs singly onto leaf hairs and approximately 30–35 eggs are laid during the whole life cycle, and these hatch after 2–3 days to give rise larvae. Young larvae have six legs, while the next two nymph stages have eight legs (like an adult). The complete transformation from an egg to an adult ordinarily takes 8–11 days at 68–77°F (20–25°C). Entirely, the mobile stages are predatory in nature, and these will forage on adults, juveniles and eggs of spider mites. As per temperatures decline, and days turn out to be smaller, in initial autumn, mites will come into a diapause so these can magnificently persist winter situations. Mites are described to become active once more as early as January, however this is influenced by food and temperatures accessibility. The healthy adult of *A. andersoni* lives for about 3 weeks and consumes various food sources [76, 77].

Predatory mite *A. andersoni* is a predator that forages on several kinds of tiny arthropods prey and pollen. It is broadly described as a predator of spider mites on fruit crops such as grapes, raspberries, peaches and apples, and many conifers. This predatory mite feeds on numerous diverse pest mites such as gall mite, russet mite and spider mite. Its foremost target pests are spider mites *Tetranychus* spp., (*T. urticae* and *T. cinnabarinus*), broad mite and cyclamen mite (*P. latus* and *P. pallidus*), European (or citrus) red mite (*Panonychus* spp.), broad mites and Eriophyid mites including the tomato rust (or russet) mite *Aculops lycopersici* Massee (Eriophyidae). These mites can also survive on young larvae of thrips, flower pollen, sugary excretions from pests and fungi, and gall midges, so, these can be introduced before the prey is present. These feed on and control all stages of phytophagous mites with all mobile life stages of *A. andersoni* to forage on prey. It is an excellent choice for pre-emergent control of pest infestations because of their varied diet and ability

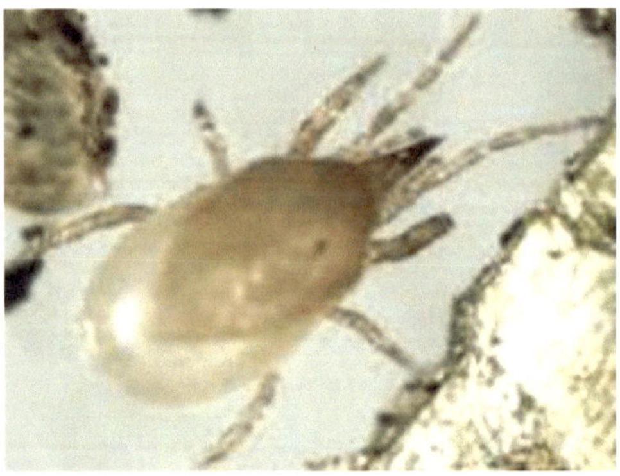

Figure 7.
Amblyseius andersoni.

to survive in cooler temperatures. These have a wide temperature tolerance of 43–104°F that makes *A. andersoni* an easy choice for preventative mite control in commercial garden [78, 79].

This predator works well on both inside protected crops and outside in ornamental crops, fruit trees, horticulture, nurseries and seedbeds. The invasion of the predator is realized under the low mass and the average density of 0.25–1 individual per 1 $m^2$of pest population in amounts. Based on the focuses of pest populations, the rate of predator application is increased. This predatory mite is suitable for biologically controlling of some mites. For scouting, if some agile-looking mites are seen running quickly across the leaf's undersurface, these are probably predators. One predatory mite per every 6 feet of crop row or 2–3 mites per 10–11 square feet are used. For best results, apply the *A. andersoni* when pest mite numbers are low or use before pest populations have reached high levels. The predatory mites will then be able to feed on small colonies of pest mites and prevent these from growing and causing major damage. Predatory mite is intended as a preventative treatment @ 1 per m^2 (curative, light @ 2–3 per m^2 at 5–7 days interval with three frequencies and curative, heavy @ 3–5 per m^2 at 5–7 days interval with three frequencies). Apply as a preventative treatment or to active mite infestations in vegetable crops, strawberries, raspberries and fruit trees, and it might be a stronger competitor under low humidity conditions [80, 81]. High numbers of this mite overwinter in the ground litter in apple orchards and by transferring such ground litter to young orchards, the population density has been found to be significantly increased in the released compared to control fields [82].

2.8 *Neoseiulus* (= *Amblyseius*) *pseudolongispinosus* (Xin, Liang and Ke)

Mite *Neoseiulus* (= *Amblyseius*) *pseudolongispinosus* (Xin, Liang and Ke) [synonym *Amblyseius* (*Neoseiulus*) *womersleyi* Schicha], is considered as the best native predator of spider mites in China. Among the stages of predator, a high predation rate has been observed in adult that showed more preference towards the eggs and nymphs of *T. urticae* [83] as well as *T. cinnabarinus* [84]. It is very similar to *Neoseiulus longispinosus* (Evans), but dorsal seta S-5 is long (as long as S-4) and barbed in *N. pseudolongispinosus* (**Figure 8**), while short (much shorter than S-4) and smooth in *N. longispinosus*. The pattern of mating behavior of *N. pseudolongispinosus* summaries that there is a period of male's climbing on the dorsum of the female. A sex attractant produced by female deutonymphs probably played a role only when the sexes physically encounter one another. The average time of a full mating lasted for 183 minutes. Most male *A. pseudolongispinosus* inseminated females with only one endospermatophore transferred to only one of the female spermathecae in single mating. The egg production and oviposition period increased with the duration of copulation. The egg production in single mating that lasted for the whole duration is only half of that in natural mating condition. The ratio of female offspring increased with the duration of copulation and then decreased slightly after 120 minutes [85–87].

Research on the biology of predator *N. pseudolongispinosus* has been carried out by usage of *T. putrescentiae*, reared on flour of soybean (*Glycine max* L.), wheat (*Triticum aestivum* L.), and maize (*Zea mays* L.), as hosts. When *N. pseudolongispinosus* nourished on *T. putrescentiae* reared on diverse foods, it could breed ordinarily and accomplished its growth as magnificently as *T. putrescentiae*. Mean period of egg and nymph development of *N. pseudolongispinosus* continued fastest statistically by means of *T. aestivum* fed *T. putrescentiae* in comparisons with *G. max* and *Z. mays* diets. Average overall longevities of male (20.8, 16.8, and 13.1 days) and the female (29.4, 25.8 and 17.7 days) have been considerably lengthier on *T. aestivum* than on *G. max* and *Z. mays* diets, correspondingly. Average fecundity of female adults

Figure 8.
Neoseiulus pseudolongispinosus.

amplified too on *T. aestivum* daily (3.5 per day), but reduced (2.8 and 1.2 per day) with *Z. mays* and *G. max*, whereas, the total growing time observed 10.1, 11.8 and 12.9 days, correspondingly. A calculation of macronutrients in wheat, soybean and maize flours displayed that wheat has additional ash and carbohydrate, and on the other hand decreased fat and protein contents, as a result, it evidenced as a prominent compound to support greater mite densities [88].

The effectiveness of hunter *N. pseudolongispinosus* as a possible biocontrol means of webworm *Loxostege sticticalis* L., has been investigated. When eggs food offered, total life cycle of predator has been accomplished on an average of 12.75 days for female or male. An average pre-oviposition time of 4.28 days noted in female, oviposition time lasted for 8.71 days, and the mean number of 1.42 eggs day^{-1} $female^{-1}$ have been observed [89].

A fenvalerate-resistant strain of *N. pseudolongispinosus* has been selected and evaluated for level of resistance. After 18 selection cycles, the LC_{50} value is 65 times higher than in wild-type strain. Genomic task showed that the resistance aspect is organized by a distinct and partly prevailing gene. The resistant strain that has an identical intrinsic rate of upsurge as the wild-type strain, might be a valuable biological control mediator of spider mites in the system of an integrated pest management (IPM) [90].

Among the effects of different developmental stages of two spider mite species, for instance, *T. urticae* and *T. cinnabarinus*, predatory mite *A. pseudolongispinosus* preferred for egg followed by nymph, while larva observed the least preferred stage. Fecundity of female fed on egg, nymph or all developmental stages has been noted significantly higher than that on larva [91]. Experiments have been carried out to observe fitness and effectiveness of Phytoseiid mites as predators of carmine spider mite *T. cinnabarinus*, greenhouse whitefly *Trialeurodes vaporariorum* (Westwood) and western flower thrips *F. occidentalis*, under greenhouse conditions in cucumber (*Cucumis sativus* L.), crop. For this trial, predatory mites *Euseius utilis* (Liang and Ke), *Euseius finlandicus* (Oudemans), *Euseius castaneae* (Wang and Xu) and *N. pseudolongispinosus* have been examined for their possibility as biological control agents in treated as well as untreated control. Among all predators, *N. pseudolongispinosus* has proved the most proficient and steadfast predator in controlling thrips and whitefly populations, contrary to *E. finlandicus* that proved better in reducing spider mite density in treated crop [92].

2.9 *Cheyletus eruditus* (Schrank)

Predatory mite *Cheyletus eruditus* (Schrank) (Acari: Cheyletidae), generally exists in bulk foodstuff storages, for instance, within warehouses. It is likewise frequently establishes in animal feedstuff, mammal and bird nests, house dust and poultry litter. The food of *C. eruditus* includes a diversity of mites and insects. Altogether, its active life stages in the absence of other prey, are cannibalistic, thus it is a predatory mite that may be used to control storage mites. It is a translucent, pale yellow in color and the male mite is slightly smaller than the female. Its characters state to categorize it as a member of the family Cheyletidae, are the occurrence of a claw on the palp tibia (forming a thumb-claw process with the tarsus) and a big gnathosoma that links the body anteriorly (rather than in the more usual slightly anteroventral position) (**Figure 9**). The two smooth curved setae and two comb-like setae on the palp tarsus are analytical characteristics of the genus *Cheyletus* [93].

The average adult's body length of *C. eruditus* (not including the mouthparts) is approximately 500 micrometers (0.5 mm) for the female (range 440–630 micrometers). The homeomorphic male is smaller, with a range of 280–320 micrometers. The heteromorphic form is about 400 micrometers long. Specific male characters are, the posterior margin of the body is more narrowly rounded than in the female, propodosomal shield is wider than in female, and gnathosoma is proportionally broader and more heavily sclerotized than in females [94].

Eggs are laid in clusters and the eggs have a gluey surface, even though these are not very definitely fixed down, but are lightly bound together with silky strands. Development and fecundity are affected by prey type and environmental situations. Per female, the total number of eggs laid ranges from about 19 to 317. The time taken for an egg to develop to an adult within the temperature range of 12–30°C, decreased with an increasing in temperature. It took 33.8 days at 18.5°C, and at 25°C acquired 15.4 days. At 76% relative humidity and with *Lepidoglyphus destructor* Schrank (grocers' itch mite) as prey, maximum longevity is 107, 67 and 56 days at 18.5, 22 and 25°C, respectively [95].

It is widespread and abundant in grain stores especially those that have significant storage mite problems. Maximum accounts of *C. eruditus* are from stored foods (commonly whole grains and flour) intended for domesticated animal and human feeding. Left over densities frequently persist in storing basins and premises afterward the substances have been removed. It as well establishes in animal bedding, nests of mammals and birds, house dust, poultry litter, and occasionally in field habitats, e.g., soil, haystacks and plant debris. For commercial use, it is reared on storage mites particularly *Acarus siro* (Linnaeus), and is the only biological control agent used to control pest mites in food storage systems, for example, in grain and

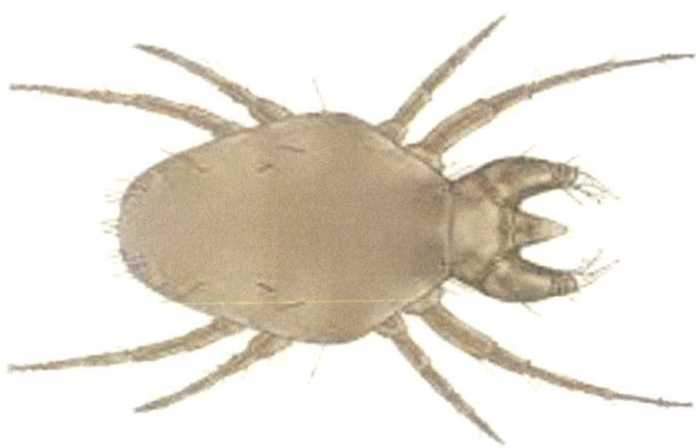

Figure 9.
Cheyletus eruditus.

seed stores. Its target pests are mainly various storage mites including the flour mite (*A. siro*)—Astigmata: Acaridae, fodder mite or grocers itch mite (*L. destructor*)—Astigmata: Glycyphagidae, and various spider mites [96].

The maximum frequently faced ectoparasite in captive snakes is the hematophagous snake mite (*Ophionyssus natricis* Gervais). Infested snakes frequently show laziness, behavioral changes (increased bathing time, rubbing against objects), pruritus, dysecdysis, crusting dermatitis (sometimes progressing to abscesses), and anemia and septicemia are special difficulties. In a study to identify the ability of the predatory mite *C. eruditus* to control *O. natricis*, 125 *O. natricis* mites have been placed in separate plastic tubes together with the same number of *C. eruditus* mites. After 48 hours, the survival rate of snake mites noted 6% compared with 92% in the control group. With another experiment, 11 infested ball pythons, with an average of 13 *O. trnaicis* per snake, have been positioned in isolated cages with 1000 *C. eruditus* mites + vermiculite. Only an average of two mites per snake persisted in comparison with 48 per snake in the control group after 15 days [97].

It can be reared in large numbers and this makes it useful in the biocontrol of pest mites that infest harvested cereal and cereal products. For bulk rearing of the predator *C. eruditus* to utilize in an extensive scale biological control of stored food mites, among several resources confirmed, lettuce seeds performed to be the best substrates for this purpose. The suggested process of bulk rearing elaborated that mite batches are retained at 25°C and R. H. 75% in paper bags on 100 g of lettuce seeds. Within 28–35 days, each batch yielded an average of 2100 ± 600 predatory mites without any additional procedures, if the initial predator-prey ratio is 1:100–1:200. The batches can either be used directly in infected stores otherwise retained at low temperature for later use [98]. The results revealed that *C. eruditus* is a mite predator with the highest natural performance in the field, indicating that it is the most pre-adapted species for biocontrol of stored-pest arthropods in grain stores [99].

3. Harmful mites

Although mites are tiny creatures, these could be extremely harmful to cause great trouble for peoples or in other ways inflicting a variety of problems associated to plants.

3.1 Spider mites (Order: Acari, Family: Tetranychidae)

Spider mites (Acari: Tetranychidae), belong to the superfamily Tetranychoidea that comprises five families, of which Tetranychidae is the largest. The common name 'spider mites' is so-called because of their ability to produce silken strands as do spiders, which is used to spin webs under that to reproduce and feed. Conversely, the silk glands in mites are situated close to mouth and allied with the mouthparts. From an applied opinion, the silk-producing habit has two vital uses for mites, firstly, falling from foliage and being adjourned from the host on a silk strand permits easy spread by wind and convection currents. Secondly, mats or tents of webbing around the mite colonies provide some degree of protection from natural enemies and treatments with pesticides.

Spider mites appear as tiny moving dots on their hosts by the naked eye. Spider mites are established on a wide variety of vascular plants, comprising shrubs, trees and herbaceous plants, from entirely all over the biosphere. Several horticultural and agricultural crops are affected by these pests, together with greenhouses and field crops, extending from low-growing bushes to fruit trees. Generally, spider mites forage on the lower side of foliage, however will cover the whole leaf surface while their densities are extraordinary. These puncture the plant cells and extract the cell contents. Their nourishing results in tiny clumps of dead cells and a spotted look of infested foliage. Wilting, leaf distortion, dryness and abscission take place with extended and high population invasions. Disturbance of photosynthesis results in plant growth checking and decrease in produce [100, 101]. Two widely distributed spider mites found on a broad range of plants are mention in the ensuing section.

3.1.1 Two-spotted spider mite Tetranychus urticae Koch

Two-spotted spider mite *Tetranychus urticae* Koch, feeds on an enormous variety of plants, including fruit trees, ornamental trees, vegetables, small fruits, shrubs and many species of weeds. The adults are typically pale green in color and adult females <0.4 mm lengthy, and in color greenish yellow having two conspicuous dark spots on either side of the back. As the mite feeds, these spots may become large enough to cover the body (**Figure 10**). Coloring and spots can be variable and may lead to misunderstanding with other mite species. Pest mite *T. urticae* reproduces through arrhenotoky, that is, a form of parthenogenesis in which unfertilized eggs develop into males. Mite overwinters as spotless, orange diapausing female beneath bark at the bottoms of trees or else in rubbish on the plantation floor. Shortly prior to bloom, these transfer to renewed shrubbery (particularly vetch and other legumes) and start nourishing on new green matter. As the climate warms and these hosts shrivel up, mites will transfer into trees, generally invading the middle portion initially. Through this time, these will have return to their characteristic green and spotted summer form and start laying eggs on the underneath of foliage [102, 103].

If a female has mated, the fertilized eggs develop into both male and female mites, if not mated, the unfertilized eggs develop into males. Eggs are shiny spheres, clear to pale green in color, pearl-like and about 0.14 mm in diameter. Eggs are laid singularly, with females depositing 5–6 eggs per day, with a total of 60–100 eggs per female. Eggs hatch in 3–6 days depending on temperature. Eggs hatch into six-legged larvae, then progress through protonymph and deutonymph stages before becoming to adults. Larvae are about the same size as eggs and the only life stage with six legs (protonymphs, deutonymphs and adults are all eight-legged). The octopods deutonymph is generally larger than the protonymph, although similar in color pattern. Larvae and nymphs complete development in 4–9 days depending on temperature and the females have a pre-oviposition period of 1–2 days. Since generations overlap, all life stages can usually be found simultaneously. There can be nine or more generations per year and adults live about 30 days [104–106].

Figure 10.
Tetranychus urticae.

Generally, the earlier a foliage is injured by mites, the more detrimental the damage will be to tree health. Midseason injury is less significant, but can combine with other stresses to cause fruit drop, poor fruit color, or reduced effectiveness of growth regulating chemicals. Some steps for spider mites controlling are scouting for the presence of pest, and noting damage and other signs of growing populations, looking for direct damage and other signs of pest, deciding if and when to take control action, choosing the best tool or tools to treat spider mites in growing situation, making spider mite treatments, and applications of biocontrol agents following the label and producer's instructions [107, 108].

One of the studies investigated the development, fecundity and population density of *T. urticae* on three different species of bean, accordingly, *Lablab purpureus* L., [Papilionaceae: Leguminosae] has been found to be a superior host plant for prey species to further culture predator *N. pseudolongispinosus* [109]. Another study examined the efficacy of four mite predators such as *E. castaneae*, *E. finlandicus*, *E. utilis* and *N. pseudolongispinosus* released for the suppression of spider mite *T. urticae* infesting sweet pepper (*C. annuum*) in greenhouse. When the predatory mites have been released, their establishment remained successful to control the population of spider mite [110].

3.1.2 Carmine spider mite Tetranychus cinnabarinus (Boisduval)

The carmine spider mite *Tetranychus cinnabarinus* (Boisduval) has the largest host range of all Tetranychidae species and is undoubtedly of greatest economic importance. The carmine spider mite has a worldwide distribution and invades approximately 100 cultured crops and weeds. It is a severe pest on eggplant, beans, pepper, cucurbits, tomatoes and various other vegetables. It is as well a pest of papaya, passion fruit and numerous further fruits. This mite also invades several flowers and ornamental plants such as chrysanthemum, carnation, cymbidium, marigold, gladiolus, rose and pikake. Nymphs and adults forage mainly on the undersides of the foliage. The upper surface of the foliage converts to stipple having the tiny dots, which are the nourishing ruptures. This pest have a habit of feeding in 'pockets' frequently adjacent to midrib and veins. Silk webbing formed by this mite is typically noticeable on host. The attacked foliage ultimately turn out to be bleached and discolored, and may drop off. The heaviest damage has been noted on glasshouse tomatoes, but the species is also frequent on other host plants such as cucumber, capsicum, aubergine and gerbera [111, 112].

The carmine spider mite is closely related to two-spotted spider mite *T. urticae* and it is difficult to distinguish both from each other in their immature stages, but its adult stage is bright red in color and more commonly found on vegetable crops than in ornamentals. Adult females are about 0.4–0.5 mm long, reddish and more or less elliptical (**Figure 11**). The males are slightly smaller than females and wedge shaped. These have a black spot on either side of their relatively colorless bodies. All stages of this mite are present throughput the year and their reproduction is most favorable when the weather is hot and dry. The adult female may live for up to 24 days and lays about 200 eggs. These eggs are laid singly on the underside of the leaf surface or attached to the silken webs spun by the adults. Eggs are spherical, shiny, straw colored and hatch in 3 days. Young larvae are a little bigger than the egg, pinkish in color and bear three pairs of legs. This stage lasts for a small period and the duration may be 1 day. There are two nymphal stages such as the protonymph and deutonymph. The nymphal stage varies from the larval stage by being somewhat bigger, having reddish or greenish color with four pairs of leg and lasts about 4 days. The carmine spider mite normally completes a life cycle from egg to adult in about 1 week [113].

Figure 11.
Tetranychus cinnabarinus.

Development times of the carmine spider mite *T. cinnabarinus* have been evaluated in the laboratory on excised leaf disc of lablab bean *Dolichos lablab* L., at 30 ± 2°C and 70 ± 5% relative humidity. Total development times from egg to adult stage have been noted 7.33 ± 0.13 days. The pre-oviposition period, oviposition period and post-oviposition period noted 0.5 ± 0, 8.05 ± 0.14 and 0.65 ± 0.07 days, respectively. Fecundity averaged 42.5 ± 1.7 eggs and longevity lasted for 9.2 ± 0.13 days [114].

The major natural predator of the carmine spider mite is a ladybird beetle *Stethorus vagans* (Blackburn) (Coccinellidae: Coleoptera), which feeds on all stages of these mites and in laboratory conditions each individual beetle consumed an average of 2400 mites [115, 116]. There are a number of predacious mites, such as *Phytoseiulus macropilis* (Banks) and several species of predatory thrips that are also effective on many crops in controlling of carmine spider mites. In orchards, it may be possible to maintain natural enemy's populations and use of sulfur to control mites in the field [117, 118].

In a study, the populations of *T. cinnabarinus* persisted greater (1.77 and 1.40 per leaf) in Bt than non-Bt varieties subsequent to insecticides usage for the control of cotton pest complex. The useful mite *N. cucumeris* persisted as vigorous in both Bt and non-Bt varieties, however Bt cotton has somewhat greater sums of the predator than non-Bt cotton (0.58 per leaf and 0.40 per leaf, respectively) to be used in creating new resistant cotton varieties as a component of an IPM strategy [119]. Biological control of the mite *T. cinnabarinus* in an open-field cotton crop (*Gossypium hirsutum* L.), by releasing the predatory mite *N. pseudolongispinosus* (Phytoseiidae) has been investigated. Field release of the predaceous mite *N. pseudolongispinosus* to reduce the incidence of *T. cinnabarinus* at an early growth stage of cotton is a potentially useful pest management strategy if every plant is treated with predator [120].

3.2 Mould mite *Tyrophagus putrescentiae* (Schrank)

Mould mite or cheese mite *Tyrophagus putrescentiae* (Schrank) in the family Acaridae, is a cosmopolitan species occurring throughout the world in a wide range of habitats, including foodstuffs, plant and animal materials, grasslands, old hay, mushrooms, and the nest of bees and ducks. It is a common pest of stored products, especially those with a high protein and fat contents including grains, nuts and seeds, and feeds on the fungi that grow on the foodstuffs [121]. The research found that *T. putrescentiae* is a fungivorous storage mite that can grow and flourish well on dog food. The study demonstrated that the presence of mould positively influences mites viability and low relative humidity can result in detrimental consequences for *T. putrescentiae* [122]. Storage mite colonizes different human-related environments and feeds on various post-harvest foods. This mite is a pest of many foods and has been found, in wheat flour, wheat germ, soy flour, inter alia, rye bread, cheese, white bread, mixtures of wheat, barley and oats, straw stacks in the field, various fruits (including dried bananas), dried milk, ham, and decaying animal and vegetable matters [123]. Mites directly endanger human health due to allergenic contamination of food, are vectors of toxicogenic fungi and thus indirectly contribute to contaminate food and feed with mycotoxins. These also cause significant grain weight losses and decrease of germination ability [124].

This mite is 0.2–0.5 mm in length and has a minute translucent body with nearly colorless legs and mouthparts. These besides have a scale on the last terminating segment of the legs. To a certain degree, their slim bodies endure a sequence of hairs, which are more frequent and lengthier than those on *A. siro*. On the underneath of the male's body, there are two dome-shaped suckers arranged on either side of the anus (**Figure 12**). The study conducted to throw some lights on the effect of some food stuffs on the main biological aspects of the Acarid mite *T. putrescentiae* at 25°C and 75% R. H., indicated that there is a slightly significant difference between the incubation period of female and food types. The females reached to maturity (life cycle) in 11.1, 12.5, 13.3 and 14.0 days when fed on wheat flour, milk powder, fish powder and granular chicken feed, respectively. Female longevity is the longest on wheat flour (39.0 days), while recorded the shortest time on fish powder (25.1 days). The female needed 34.6, 30.4, 22.8 and 21.0 days for egg deposition, respectively. The highest number of deposited eggs is recorded on wheat flour (39.0 eggs), while the lowest with fish powder (27.8 eggs) as feeding source. However, the male life cycle duration is significantly affected by the food type, as it averaged 9.8, 10.3, 12.4 and 13.2 days on the aforementioned diets, respectively. Male adulthood is the longest on wheat flour (28.7 days), while decreased to 17.0 days on fish powder. The male life span lasted for 38.5, 34.2, 31.2 and 30.2 days, when fed on the above mentioned diets, respectively [125].

Under optimum conditions, a generation can be completed in 8–21 days. As the temperature falls, the length of the life cycle increases greatly. This mite will tolerate high temperatures, and the larval stage is particularly susceptible to low and high

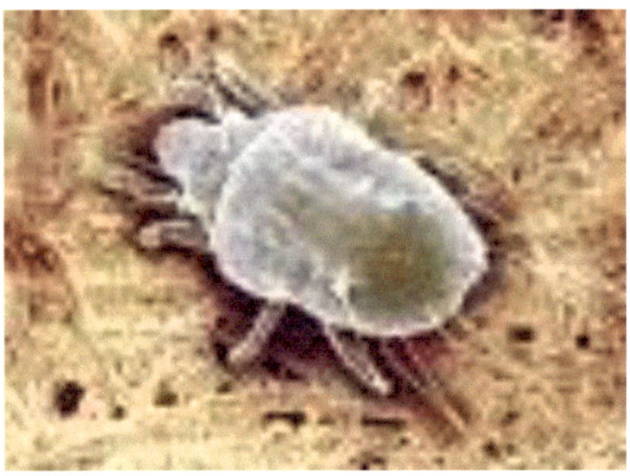

Figure 12.
Tyrophagus putrescentiae.

temperatures with 93.6 and 54% mortality at 10 and 34°C, respectively. Unlike *A. siro*, it does not produce a hypopus stage. Under ideal conditions, with temperatures above 30°C (86°F) and humidity above 85%, it can complete its life cycle in under 3 weeks [126].

Laboratory investigations on the biology of *T. putrescentiae* have been conducted using the flour of soybean (*G. max*), wheat (*T. aestivum*) and maize (*Z. mays*) as hosts. Life stages of *T. putrescentiae* (egg, larval, protonymphal and deutonymphal) have faster development rate that fed on *T. aestivum* compared to those nourished on *Z. mays* and *G. max*. Higher fecundity of *T. putrescentiae* has been observed (23.8 eggs) on *T. aestivum*, in comparison to the other diets experienced (17.1 eggs on *Z. mays* and 11.4 with *G. max*), longevity of female averaged 34.1, 27.0 and 40.8 days, whereas male longevity distinguished as 23.5, 18.7 and 28.7 days when reared on *Z. mays*, *G. max* and *T. aestivum*, respectively. On a food of *T. aestivum*, the average generation period of *T. putrescentiae* from egg to adult lasted for 11.7 days corresponding to 15.2 and 18.8 days after served with *Z. mays* and *G. max*, and net population growth known as 119.0, 61.0 and 32.0 mites per gm of substrate, respectively [88].

An IPM strategy has been developed to manage infestations of mould mite in stored animal feed, due to the increasing importance as pest of storage facilities and feed processing. This approach includes some features such as adopting striking hygiene practice in and around the processing and storage facility, controlling the moistness content of the processed feed to 12%, rejection of infested grain at the receiving point, and admixing vegetable oil to some feed (2% w/w). Moreover, seven contact insecticides and phosphine fumigant for their effectiveness against the mould mite have been evaluated to measure their potential integration into the IPM tactic. Amongst these, pyrethrin synergized with a newly developed bacterium-based material spinosad, piperonyl butoxide and insect growth regulator s-methoprene controlled the mites. Moreover, the fumigant phosphine at 1 mg/L over a 6 days exposure period also controlled these mites. Until now, the IPM tactic, has resulted in a complete eradication of the mite population in this particular case of stored animal feed [127].

Even though, the predatory mites aggressively feed on many pest species, their reproduction and dispersion to cover the affected area and time spent in prey searching can slow the mites management. Because of this limitation, *N. cucumeris* is commonly used as a precautionary controller device and can deliver a competent control of pest in its initial stage of invasion. This predator can be combined with some of the chemical insecticides (flonicamid, buprofezin, fenoxycarb, pymetrozine) for the ecologically managing of several pests. Conversely, chemical like bifenthrin, abamectin, cyfluthrin, acephate, esfenvalerate, chlorfenapyr, spinosad, fipronil, thiamethoxam and imidacloprid can be lethal to life stages of *N. cucumeris*. Within a production system, a planned rotation of *N. cucumeris* by benign insecticides can decrease the general usage of severe chemistries and interrupt the insecticide resistance selection in a pest population [128].

In the case of multiple pests inhabiting different plant parts, a higher rate or multiple predator releases may be required to achieve the desired level of control. In some natural pest control programs, various predator species are released to manage a single prey species. Whereas, in some circumstances, release of multiple species may offer a well control, while in other cases species may interact with each other for a possible negative outcome on biological control package. In a study, intraguild predation has been evaluated between three phytoseiid species, *P. persimilis*, *A. swirskii* and *N. barkeri* in laboratory situations in either presence or absence of *T. urticae* and/or pollen. Adult females of entire three predator species revealed higher predation rates on larvae than on the other immature stages. Mite predator *P. persimilis* did not forage on the nymphal stages of the other two phytoseiid species, whereas *A. swirskii* and *N. barkeri* fed on all juvenile stages of the two others. Females of *A. swirskii* devoured more phytoseiid larvae than did the other two species. On the other hand, the predation of the three female species on immature stages reduced considerably when prey/food has been added to experimental units. Results advocate that three species, *P. persimilis*, *A. swirskii* and *N. barkeri* are possibly prone to intraguild interactions with each other, and *A. swirskii* is the durable intraguild predator. The outcomes of this study may be supportive in choosing of effective biological control approaches against spider mites [129].

Predator mites are most effective when applied at the first sign of a mites or insect pests infestations. These will typically turn out to be established in the crop afterward one introduction, wherever there persist either mites or pollen for diet. When prey become infrequent, for example, *N. fallacis*, transfers to the upper portion of the plant and commonly scatters all over the crop by the wind or on air currents. Once predators are set up on some infested foliage, it is generally meaning that the biological control package will be fruitful. It may take another 2–6 weeks for new plant growth to display improvement, reliant on growth degrees. For best results in field crops, placing higher numbers of predators on the prevailing upwind side of the crop will increase their dispersal throughout the crop via wind. In greenhouses, natural enemy Persimilis should always be applied along with Fallacis predator [130].

4. Conclusion

Ongoing studies include the biology and ecology of some mite predators along with pest mites and how biotic and abiotic factors affect pests and their natural enemies. To cut a long story into short, from lookout of biological control of pest mites or insect pests, the knowledge on biology and ecology of some predacious as well as harmful mites is undoubtedly important. A successful management plan requires information about a species biology including its diet, lifecycle and mass releases of predator, how it interacts with the environment and with other species as well as species behavior and how the behavior of both pests and beneficial enemies can be manipulated to reduce or prevent yield losses. Information of the biology and ecology of mite pests and their natural enemies contained in the chapter is a prerequisite to keep a minimum economic impact of pests, eliminate pest menaces by organic pest controlling and implement efficient plantation protection practices with modern thinking on environmental problems.

Author details

Muhammad Sarwar
National Institute for Biotechnology and Genetic Engineering (NIBGE), Faisalabad, Pakistan

*Address all correspondence to: drmsarwar64@gmail.com

References

[1] Sarwar M. Diseases transmitted by blood sucking mites and integrated mite management for their prevention. American Journal of Food Science and Health. 2016;**2**(6):169-175

[2] Sarwar M. Mites (Arachnida: Acarina) affecting humans and steps taking for the solution of problematics. International Journal for Research in Mechanical Engineering. 2016;**1**(7):1-14

[3] Sarwar M. Mites (Acarina) as vectors of plant pathogens and relation of these pests to plant diseases. Agricultural and Biological Sciences Journal. 2015;**1**(4):150-156

[4] Ashfaq M, Sarwar M, Pervez A. A description of two new species (hypopi) of the genus *Acotyledon* (Acariformes: Acaridae) from Pakistan. Russian Journal of Acarology, Acarina. 1998;**6**(1-2):51-57

[5] Sarwar M, Ashfaq M, Aslam M. Two new species of genus *Histiostoma* from Pakistan (Acari: Histiostomatidae). Pakistan Entomologist. 1998;**20**(1-2):1-10

[6] Ashfaq M, Sarwar M, Aslam M. Taxonomic studies on hypopi of genus *Acotyledon* Oudemans from Pakistan. Pakistan Entomologist. 1998;**20**(1-2):66-74

[7] Ashfaq M, Sarwar M, Akbar S. A new species (hypopus) of the genus *Lackerbaueria* Zakhvatkin (Acarina: Acaridae) from Pakistan. Russian Journal of Acarology, Acarina. 1999;7(1):53-56

[8] Ashfaq M, Sarwar M. Three new hypopi of genus *Acotyledon* Oudemans (Acarina: Acaridae) from Pakistan. Pakistan Entomologist. 1999;**21**(1-2):3-14

[9] Ashfaq M, Sarwar M, Akbar S. Two new (hypopus) of the genus *Glyphanoetus* Oudemans (Acari: Histiostomatidae). Journal of Acarological Society of Japan. 1999;**8**(1):21-26

[10] Sarwar M. New mite (hypopus) from central Punjab (Pakistan) in the family Acaridae (Acari: Astigmata). Pakistan Journal of Scientific and Industrial Research. 2000;**43**(6):374-377

[11] Ashfaq M, Sarwar M, Ali A. Two new mite species (hypopi) of the genus *Histiostoma* (Acari: Histiostomatidae) from Pakistan. Pakistan Journal of Agricultural Sciences. 2000;**37**(1-2):33-41

[12] Sarwar M. *Forcellinia tumulus*, a new species in Pakistan belonging to the family Acaridae (Acari.). Pakistan Journal of Scientific and Industrial Research. 2001;**44**(1):33-37

[13] Ashfaq M, Sarwar M. Description of two new species (hypopus) of the genus *Copronomoia* Mahunka (Acari: Histiostomatidae) from Pakistan. Russian Journal of Acarology, Acarina. 2001;**9**(1):125-130

[14] Sarwar M, Ashfaq M. Contribution towards the taxonomic study on the mite (Hypopus) of the genus *Histiostoma* Oudemans (Acarina: Histiostomatidae) of Pakistan. Balochistan Journal of Agricultural Sciences. 2002;**3**(2):25-35

[15] Siddiqui QH, Sarwar M. Pre and post-harvest losses in wheat. Pakistan and Gulf Economist. 2002;**21**(6):30-32

[16] Sarwar M, Ashfaq M. Two new *Caloglyphus* Berlese mites (Astigmata: Acaridae) recorded in Pakistan. Pakistan Journal of Scientific and Industrial Research. 2004;**47**(6):455-461

[17] Sarwar M. Stored grain and stored product mites from Pakistan and Azad Kashmir. Pakistan and Gulf Economists. 2004;**23**(10):30-31

[18] Sarwar M, Ashfaq M, Akbar S. Numerical taxonomy of two new mite species of the genus *Caloglyphus* Berlese (Acaridae) from Pakistan. Pakistan Journal of Scientific and Industrial Research. 2005;**48**(5):345-353

[19] Sarwar M, Ashfaq M, Nadeem S. On the identity of new acarid mites in genus *Caloglyphus* Berlese occurring in Asian expanse (Pakistan) (Acarina: Acaridae). Journal of Agriculture and Biological Sciences. 2009;**1**(1):38-47

[20] Sarwar M, Ashfaq M. Contribution toward the description of a new *Caloglyphus* Berlese mite (Acarina: Acaridae) from collections in Pakistan. Biological Diversity and Conservation. 2010;**3**(1):45-50

[21] Sarwar M, Ashfaq M. Description of a new *Caloglyphus* Berlese mite (Acari: Acaridae) infesting pulse in Pakistan. The Nucleus. 2010;**47**(1):61-66

[22] Sarwar M, Ashfaq M. Studies on taxonomy of a new mite species in the genus *Caloglyphus* Berlese (Acarina: Acaridae) recognized in Pakistan. Journal of Cereals and Oilseeds. 2012;**3**(2):15-20

[23] Sarwar M, Ashfaq M. An undescribed mite of genus *Caloglyphus* Berlese (Acarina: Acaridae) collected from oriental region, Pakistan. Biological Diversity and Conservation. 2012;**5**(2):1-6

[24] Sarwar M. Feasibility for development of comparative life histories and predation of predatory mites in Phytoseiidae complex and their experimental manipulations for pests control. International Journal of Animal Biology. 2015;**1**(5):150-157

[25] Sarwar M. Mite pests (Acari) in mango (*Mangifera indica* L.) plantations and implementation of control strategy. Bioscience and Bioengineering. 2015;**1**(3):41-47

[26] Sarwar M. Frequency of insect and mite Fauna in chilies *Capsicum annum* L., onion *Allium cepa* L. and garlic *Allium sativum* L. cultivated areas, and their integrated management. International Journal of Agronomy and Plant Production. 2012;**3**(5):173-178

[27] Sarwar M, Ashfaq M, Ahmad A, Randhawa MA. Assessing the potential of assorted plant powders on survival of *Caloglyphus* grain mite (Acari: Acaridae) in wheat grain. International Journal of Agricultural Science and Bioresource Engineering Research. 2013;**2**(1):1-6

[28] Sarwar M. Studies on resistance discrepancy in seeds of local and exotic lentil (*Lens culinaris* Medikus) genotypes against *Callosobruchus analis* L. (Coleoptera: Bruchidae). African Entomology: Journal of the Entomological Society of Southern Africa. 2013;**21**(2):273-279

[29] Sarwar M, Saqib SM. Rearing of predatory seven spotted ladybird beetle *Coccinella septempunctata* L. (Coccinellidae) on natural and artificial diets under laboratory conditions. Pakistan Journal of Zoology. 2010;**42**(1):47-51

[30] Shah SMM, Ahmad N, Sarwar M, Tofique M. Rearing of *Bactrocera zonata* (Diptera: Tephritidae) for parasitoids production and managing techniques for fruit flies in mango orchards. International Journal of Tropical Insect Science. 2014;**34**(S1):108-113

[31] Sarwar M, Ahmad N, Rashid A, Shah SMM. Valuation of gamma irradiation for proficient production of parasitoids (Hymenoptera: Chalcididae & Eucoilidae) in the management of the peach fruit-fly, *Bactrocera zonata* (Saunders). International Journal of Pest Management. 2015;**61**(2):126-134

[32] Grafton-Cardwell EE, Ouyang Y, Striggow RA. Predacious mites (Acari: Phytoseiidae) for control of spider

mites (Acari: Tetranychidae) in nursery citrus. Environmental Entomology. 1997;**26**:121-130

[33] McMurtry JA, Croft BA. Life-styles of phytoseiid mites and their roles in biological control. Annual Review of Entomology. 1997;**42**:291-321

[34] Moraes GJD, McMurtry JA, Denmark HA, Campos CB. A revised catalog of the mite family Phytoseiidae. Zootaxa. 2004;**434**:1-494

[35] Beard JJ. Taxonomy and biological control: *Neoseiulus cucumeris* (Acari: Phytoseiidae), a case study. Australian Journal of Entomology. 1999;**38**:51-59

[36] Sarwar M. Comparative life history characteristics of the mite predator *Neoseiulus cucumeris* (Oudemans) (Acari: Phytoseiidae) on mite and pollen diets. International Journal of Pest Management. 2016;**62**:140-148

[37] Sarwar M, Kongming W, Xuenong X. Evaluation of biological aspects of the predacious mite, *Neoseiulus cucumeris* (Oudemans) (Acari: Phytoseiidae) due to prey changes using some selected arthropods. International Journal of Acarology. 2009;**35**(6):503-509

[38] Brodsgaard HF, Hansen LS. Effect of *Amblyseius cucumeris* and *Amblyseius barkeri* as biological control agents of *Thrips tabaci* on glasshouse cucumbers. Biocontrol Science and Technology. 1992;**2**:215-223

[39] Brodeur J, Bouchard A, Turcotte G. Potential of four species of predatory mites as biological control agents of the tomato russet mite, *Aculops lycopersici* (Massee) (Eriophyidae). Canadian Entomologist. 1997;**129**:1-6

[40] Weintraub PG, Kleitman S, Mori R, Shapira N, Palevsky E. Control of broad mites (*Polyphagotarsonemus latus* (Banks)) on organic greenhouse sweet peppers (*Capsicum annuum* L.) with the predatory mite, *Neoseiulus cucumeris* (Oudemans). Biological Control. 2003;**26**:300-309

[41] Arthurs S, McKenzie CL, Chen J, Dogramaci M, Brennan M, Houben K, et al. Evaluation of *Neoseiulus cucumeris* and *Amblyseius swirskii* (Acari: Phytoseiidae) as biological control agents of chili thrips, *Scirtothrips dorsalis* (Thysanoptera: Thripidae) on pepper. Biological Control. 2009;**49**:91-96

[42] Kakkar G, Kumar V, Seal DR, Liburd OE, Stansly P. Predation by *Neoseiulus cucumeris* and *Amblyseius swirskii* on *Thrips palmi* and *Frankliniella schultzei* on cucumber. Biological Control. 2016;**92**(1):85-91

[43] Navajas M, Thistlewood H, Lagnel J, Marshall D, Tsagkarakou A, Pasteu N. Field releases of the predatory mite *Neoseiulus fallacis* (Acari: Phytoseiidae) in Canada, monitored by pyrethroid resistance and allozyme markers. Biological Control. 2001;**20**(3):191-198

[44] Helle W, Sabelis MW. Spider Mites: Their Biology, Natural Enemies and Control. Vol. 1B. Amsterdam: Elsevier; 1985. p. 458

[45] Haralabos T, Salvatore R. On the identity of *Neoseiulus fallacis* (Garman 1948) (Parasitiformes, Phytoseiidae): Redescription of the species and description of the new species *Neoseiulus garmani*. International Journal of Acarology. 2016;**42**(8):394-404

[46] Croft BA, Messing RH, Dunley JE, Strong WB. Effects of humidity on eggs and immatures of *Neoseiulus fallacis, Amblysieus andersoni, Metaseiulus occidentalis* and *Typhlodromus pyri* (Phytoseiidae): Implications for biological control on apple, caneberry, strawberry and hop. Experimental and Applied Acarology. 1993;**17**(6):451-459

[47] Halliday RB. Mites of Australia: A checklist and bibliography. In:

Monographs on Invertebrate Taxonomy, Vol. 5. Australia: CSIRO; 1998. p. 317

[48] Pratt PD, Rosetta R, Croft BA. Plant-related factors influence the effectiveness of *Neoseiulus fallacis* (Acari: Phytoseiidae), a biological control agent of spider mites on landscape ornamental plants. Journal of Economic Entomology. 2002;**95**:1135-1141

[49] Naher L, Haque M. Biological control of *Tetranychus urticae* (Acari: Tetranychidae) using *Phytoseiullus persimilis* (Acari: Phytoseiidae). Research Journal of Agriculture and Biological Sciences. 2007;**3**:550-553

[50] Ali FS. Life tables of *Phytoseilus macropilis* (Banks) (Gamasida: Phytseiidae) at different temperatures. Experimental and Applied Acarology. 1998;**22**:335-342

[51] Escudero LA, Ferragut F. Life-history of predatory mites *Neoseiulus californicus* and *Phytoseiulus persimilis* (Acari: Phytoseiidae) on four spider mite species as prey, with special reference to *Tetranychus evansi* (Acari: Tetranychidae). Biological Control. 2005;**32**:378-384

[52] Canias L, Amano H, Ochiai N, Takeda M. Biology and predation of the Japanese strain of *Neoseiulus californicus* (McGregor) (Acari: Phytoseiidae). Systematic and Applied Acarology. 2006;**11**:141-157

[53] Tello V, Vargas R, Araya J, Cardemil A. Biological parameters of *Cydnodromus picanus* and *Phytoseiulus persimilis* raised on the carmine spider mite, *Tetranychus cinnabarinus* (Acari: Phytoseiidae, Tetranychidae). Ciencia e Investigación Agraria. 2009;**36**(2):277-290

[54] Hoffmann MP, Frodsham AC. Natural Enemies of Vegetable Insect Pests. Ithaca, NY: Cooperative Extension, Cornell University; 1993. p. 63

[55] Opit GP, Nechols JR, Margolies DC. Biological control of two spotted spider mites, *Tetranychus urticae* Koch (Acari: Tetranychidae), using *Phytoseiulus persimilis* Athias-Henriot (Acari: Phytoseidae) on ivy geranium: Assessment of predator release ratios. Biological Control. 2004;**29**:445-452

[56] Tixier MS, Guichou S, Kreiter S. Morphological variation in the biological control agent *Neoseiulus californicus* (McGregor) (Acari: Phytoseiidae): Consequences for diagnostic reliability and synonymies. Invertebrate Systematics. 2008;**22**:453-469

[57] Gotoh L, Tsuchiya A, Kitashima Y. Influence of prey on developmental performance, reproduction and prey consumption of *Neoseiulus californicus* (Acari: Phytoseiidae). Experimental and Applied Acarology. 2006;**40**:189-204

[58] Greco NM, Sanchez NE, Liljesthrom GG. *Neoseiulus californicus* (Acari: Phytoseiidae) as a potential control agent of *Tetranychus urticae* (Acari: Tetranychidae): Effect of pest/predator ratio on pest abundance on strawberry. Experimental and Applied Acarology. 2005;**37**(1-2):57-66

[59] Fraulo AB, Liburd OE. Biological control of twospotted spider mite, *Tetranychus urticae*, with predatory mite, *Neoseiulus californicus*, in strawberries. Experimental and Applied Acarology. 2007;**43**:109-119

[60] Xu X, Wang B, Wang E, Zhang ZQ. Comments on the identity of *Neoseiulus californicus* sensu lato (Acari: Phytoseiidae) with a redescription of this species from southern China. Systematic and Applied Acarology. 2013;**18**:329-344

[61] Calvo FJ, Bolckmans K, Belda JE. Control of *Bemisia tabaci* and

Frankliniella occidentalis in cucumber by *Amblyseius swirskii*. BioControl. 2011;**56**:185-192

[62] Stansly PA, Castillo JA. Control of broadmites, spidermites, and whiteflies using predaceous mites in open-field pepper and eggplant. Proceedings of the Florida State Horticultural Society. 2010;**122**:253-257

[63] Onzo A, Houedokoho AF, Hanna R. Potential of the predatory mite, *Amblyseius swirskii* to suppress the broad mite, *Polyphagotarsonemus latus* on the gboma eggplant, *Solanum macrocarpon*. Journal of Insect Science. 2012;**12**:7

[64] Juan-Blasco M, Qureshi JA, Urbaneja A, Stansley PA. Predatory mite, *Amblyseius swirskii* (Acari: Phytoseiidae), for biological control of Asian citrus psyllid, *Diaphorina citri* (Hemiptera: Psyllidae). Florida Entomologist. 2012;**95**:543-551

[65] Messelink G. *Typhlodromips swirskii* (Athias-Henriot) (Acari: Phytoseiidae): A new predator for thrips control in greenhouse cucumber. Integrated Control in Protected Crops, Temperate Climate IOBC/WPRS Bulletin. 2005;**28**(1):183-186

[66] Chen X, Zhang Y, Ji J, Lin J. Experimental life table for population of *Amblyseius swirskii* (Athias-Henriot) fed on *Tetranychus truncatus* (Ehara). Fujian Journal of Agricultural Sciences. 2011;**3**:018

[67] Lee HS, Gillespie DR. Life tables and development of *Amblyseius swirskii* (Acari: Phytoseiidae) at different temperatures. Experimental and Applied Acarology. 2011;**53**:17-27

[68] Park HH, Shipp L, Buitenhuis R. Predation, development and oviposition by the predatory mite *Amblyseius swirskii* (Acari: Phytoseiidae) on tomato russet mite (Acari: Eriophyidae). Journal of Economic Entomology. 2010;**103**:563-569

[69] Park HH, Shipp L, Buitenhuis R, Ahn JJ. Life history parameters of a commercially available *Amblyseius swirskii* (Acari: Phytoseiidae) fed on cattail (*Typha latifolia*) pollen and tomato russet mite (*Aculops lycopersici*). Journal of Asia-Pacific Entomology. 2011;**14**:497-501

[70] Buitenhuis R, Shipp L, Scott-Dupree C. Dispersal of *Amblyseius swirskii* Athias-Henriot (Acari: Phytoseiidae) on potted greenhouse chrysanthemum. Biological Control. 2010;**52**:110-114

[71] Opit GP, Nechols JR, Margolies DC, Williams KA. Survival, horizontal distribution, and economics of releasing predatory mites (Acari: Phytoseiidae) using mechanical blowers. Biological Control. 2005;**33**:344-351

[72] Hoy CW, Glenister CS. Releasing *Amblyseius* spp. (Acarina: Phytoseiidae) to control *Thrips tabaci* (Thysanoptra: Thripidae) on cabbage. Entomophaga. 1991;**34**(6):561-573

[73] Bonde J. Biological studies including population growth parameters of the predatory mite *Amblyseius barkeri* [Acarina: Phytoseiidae] at 25°C in the laboratory. Entomophaga. 1989;**34**(2):275-287

[74] Reda AS, Momen FM. Influence of prey availability on development, prey consumption and reproduction of *Typhlodromus pelargonicus* and *Amblyseius barkeri* (Acarina: Phytoseiidae). Bulletin of the Institute of Zoology Egypt. 1993;**41**:105-114

[75] Momen FM. Feeding, development and reproduction of *Amblyseius barkeri* (Acarina: Phytoseiidae) on various kinds of food substances. Acarologia. 1995;**36**(2):101-105

[76] Croft BA, Croft MB. Larval survival and feeding by immature *Metaseiulus occidentalis*, *Amblyseius andersoni* and *Typhlodromus pyri* on life stage

groups of *Neoseiulus fallacis Tetranychus urticae* Koch and phytoseiid larvae. Experimental and Applied Acarology. 1993;**17**(9):685-693

[77] Zhang ZQ, Croft BA. A comparative life history study of immature *Amblyseius fallacies*, *Amblyseius andersoni*, *Typhlodromus occidentalis* and *Typhlodromus pyri* (Acari: Phytoseiidae) with a review of larval feeding patterns. Experimental and Applied Acarology. 1994;**18**:631-657

[78] Ivancich GP. Natural alternative food for *Amblyseius andersoni* Chant (Acarina: Phytoseiidae) on plants without prey. Redia. 1988;**71**:161-171

[79] Duso C, Camporese P, Van Der Geest LPS. Toxicity of a number of pesticides to strains of *Typhlodromus pyri* and *Amblyseius andersoni* (Acari: Phytoseiidae). Entomophaga. 1992;**37**:363-372

[80] Duso C, Pasini M. Distribution of the predatory mite *Amblyseius andersoni* (Acari: Phytoseiidae) on different apple cultivars. Anz. Schadlingsk. Journal of Pest Science. 2003;**76**:33-40

[81] Chant DA, JA MM. A review of the subfamily Amblyseiinae Muma (Acari: Phytoseiidae): Part I. Neoseiulini new tribe. International Journal of Acarology. **29**(1):3-46

[82] Szabo P, Arpad B. A new method for the release of *Amblyseius andersoni* (Acari: Phytoseiidae) in young apple orchards. European Journal of Entomology. 2013;**110**(3):477-482

[83] Zou P, Gao JR, Ma EP. A study of food preference and predatory capacity of *Amblyseius pseudolongispinosus*. Natural Enemies of Insects. 1986;**8**(3):137-141

[84] Jiang HL, Wang ED, Lu JL, Wang BM, Xu XN. Preference of *Neoseiulus californicus* (Acari: Phytoseiidae) and functional responses of *N. californicus* and *Amblyseius pseudolongispinosus* to prey developmental stages of *Tetranychus cinnabarinus*. Chinese Journal of Biological Control. 2015;**31**(1):8-13

[85] Xin JL, Liang LR, Ke LS. A new species of the genus *Amblyseius* from China (Acarina: Phytoseiidae). International Journal of Acarology. 1981;**7**(1-4):75-80

[86] Zhou AN, Chang SS. Experimental studies on the biological characteristics of *Amblyseius pseudolongispinosus* (Acari: Phytoseiidae). Chinese Journal of Biological Control. 1989;**5**(4):153-156

[87] Xu X, Liang L. Studies on the behavior and reproduction of *Amblyseius pseudolongispinosus* (Xin, Liang & Ke) (Acarina: Phytoseiidae). Journal of Anhui Agricultural University. 1994;**21**:81-86

[88] Sarwar M, Xuenong X, Kongming W. Effects of different flours on the biology of the pry *Tyrophagus putrescentiae* (Schrank) (Acarina: Acaridae) and the predator *Neoseiulus pseudolongispinosus* (Xin, Liang and Ke) (Acari: Phytoseiidae). International Journal of Acarology. 2010;**36**(5):363-369

[89] Sarwar M, Xuenong X, Kongming W. Suitability of webworm *Loxostege sticticalis* L. (Lepidoptera: Crambidae) eggs for consumption by immature and adults of the predatory mite *Neoseiulus pseudolongispinosus* (Xin, Liang and Ke) (Acarina: Phytoseiidae). Spanish Journal of Agricultural Research. 2012;**10**(3):786-793

[90] Zhu K, Lisheng K, Jieliu X. Selection and genetic analysis of fenvalerate resistance in *Amblyseius pseudolongispinosus* (Acari: Phytoseiidae). Systematic and Applied Acarology. 1996;**1**(1):5-10

[91] Sarwar M, Xuenong X, Wang E, Kongming W. The potential of four mite species (Acari: Phytoseiidae) as predators of sucking pests on protected cucumber (*Cucumis sativus* L.) crop. African Journal of Agricultural Research. 2011;**6**(1):73-78

[92] Yang Y, Jiang XH, Xu XN, Wang ED. Effects of two species of spider mites and their different developmental stages on the offspring sex ratios of the predatory mite *Amblyseius pseudolongispinosus* (Acari: Phytoseiidae). Chinese Journal of Biological Control. 2013;**29**(1):17-23

[93] Barker PS. Bionomics of *Cheyletus eruditus* (Schrank) (Acarina: Cheyletidae), a predator of *Lepidoglyphus destructor* (Schrank) (Acarina: Glycyphagidae). Canadian Journal of Zoology. 1991;**69**(9):2321-2325

[94] Fain A, Bochkov AV. A review of the genus *Cheyletus* Latreille, 1776 (Acari: Cheyletidae). Bulletin de l'Institut Royal des Sciences Naturelles de Belgique, Entomologie. 2001;**71**:83-113

[95] Zdarkova E, Horak P. Development of *Cheyletus eruditus* (Schrank) (Acarina: Cheyletidae) at low temperature. Plant Protection Science. 1999;**35**:14-16

[96] Pekar S, Zdarkova E. A model of the biological control of *Acarus siro* by *Cheyletus eruditus* (Acari: Acaridae, Cheyletidae) on grain. Journal of Pest Science. 2004;**77**(1):1-10

[97] Schilliger LH, Morel D, Bonwitt JH, Marquis O. *Cheyletus eruditus* (taurrus): An effective candidate for the biological control of the snake mite (*Ophionyssus natricis*). Crop Protection. 1986;**5**(2):122-124

[98] Zdarkova E. Mass rearing of the predator *Cheyletus eruditus* (Schrank) (Acarina: Cheyletidae) for biological control of acarid mites infesting stored products. Journal of Zoo and Wildlife Medicine. 2013;**44**(3):654-659

[99] Lukas J, Stejskal V, Jarosik V, Hubert J, Zdarkova E. Differential natural performance of four *Cheyletus* predatory mite species in Czech grain stores. Journal of Stored Products Research. 2007;**43**:97-102

[100] Sabelis MW. Development. In: Helle W, Sabelis MW, editors. Spider Mites, their Biology, Natural Enemies and Control. World Crop Pests. Vol. IB. Amsterdam, The Netherland: Elsevier; 1985. pp. 43-53

[101] Sarwar M. Some insect pests (Arthropoda: Insecta) of summer vegetables, their identification, occurrence, damage and adoption of management practices. International Journal of Sustainable Agricultural Research. 2014;**1**(4):108-117

[102] Raworth DA, Gillespie DR, Roy M, Thistlewood HMA. *Tetranychus urticae* Koch, two spotted spider mite (Acari: Tetranychidae). In: Mason PG, Huber JT, editors. Biological Control Programmes in Canada, 1981-2000. CAB International; 2002. pp. 259-265

[103] Sarwar M. Knowing about identify and mode of damage by insect pests attacking winter vegetables and their management. Journal of Ecology and Environmental Sciences. 2014;**2**(4):1-8

[104] Enohara K, Amano H. Simple method for discriminating six common species of red *Tetranychus* spider mites (Acari: Tetranychidae) in Japan. Japanese Journal of Applied Entomology and Zoology. 1996;**40**(4):311-315

[105] Zhang ZQ, Jacobson RJ. Using adult female morphological characters for differentiating *Tetranychus urticae* complex (Acari: Tetranychidae) from greenhouse tomato crops in UK. Systematic and Applied Acarology. 2000;**5**:69-76

[106] Elmoghazy MME. Survey and taxonomy of mites associated with fruit orchards trees from Sakaka governorate, Kingdom of Saudi Arabia. International Journal of Agriculture and Biosciences. 2016;**5**(6):341-346

[107] Watanabe MA, Moraes G, Gastaldo IJ, Nicolella G. Biological control of two spotted spider mite with predatory phytoseiids (Acari: Tetranychidae, Phytoseiidae) on cucumber and strawberry. Scientia Agricola. 1994;**51**(1):75-81

[108] Zhang ZQ. Mites of Greenhouses: Identification, Biology and Control. Wallingford, UK: CAB International; 2003. p. 244

[109] Sarwar M. Influence of host plant species on the development, fecundity and population density of pest *Tetranychus urticae* Koch (Acari: Tetranychidae) and predator *Neoseiulus pseudolongispinosus* (Xin, Liang and Ke) (Acari: Phytoseiidae). New Zealand Journal of Crop and Horticultural Science. 2014;**42**(1):10-20

[110] Sarwar M, Kongming W, Xuenong X, Wang E. Evaluations of four mite predators (Acari: Phytosiidae) released for suppression of spider mite infesting protected crop of sweet pepper (*Capsicum annuum* L.). African Journal of Agricultural Research. 2011;**6**(15):3509-3514

[111] Popov SY. Long-term characteristics of seasonal development of spider mites of the genus *Tetranychus* Dufour, 1832 (Acariformes, Tetranychidae) on strawberry in Moscow Province. Entomologicheskoe Obozrenie. 2003;**82**(1):71-85

[112] Ghoshal S, Gupta SK, Mukherjee B. Seasonal abundance of phytophagous and predatory mites on mangrove vegetation and agri-horticultural crops of Sundarban biosphere reserve. Acarina. 2004;**12**(1):49-56

[113] Ashutosh S, Pratap KP. First record of the carmine spider mite, *Tetranychus urticae*, infesting *Withania somnifera* in India. Journal of Insect Science. 2012;**12**:50

[114] Kaimal SG, Ramani N. Life cycle of *Tetranychus cinnabarinus* (Boisduval) (Acari: Tetranychidae) on Lablab bean. Indian Journal of Fundamental and Applied Life Sciences. 2011;**1**(2):43-47

[115] Sarwar M. Recognition of some lady beetles (Coleoptera: Coccinellidae) deadly sighted for insect and mite pests in agroecosystems. International Journal of Entomology Research. 2016;**1**(2):29-34

[116] Sarwar M. Biological control to maintain natural densities of insects and mites by field releases of lady beetles (Coleoptera: Coccinellidae). International Journal of Entomology and Nematology. 2016;**2**(1):21-26

[117] Sarwar M. Mites—the tiny killers to push honeybee colonies into collapse and integrated pest management. International Journal for Research in Applied Physics. 2016;**1**(7):12-21

[118] Sarwar M. Mite culprits for causing mortality and reduction in population of honey bee colonies and measures for pests control. International Journal for Research in Applied Chemistry. 2016;**1**(7):10-22

[119] Sarwar M. Comparing abundance of predacious and phytophagous mites (Acarina) in conjunction with resistance identification between Bt and non-Bt cotton cultivars. African Entomology: Journal of the Entomological Society of Southern Africa. 2013;**21**(1):108-118

[120] Sarwar M. Management of spider mite *Tetranychus cinnabarinus* (Boisduval) (Tetranychidae) infestation in cotton by releasing the predatory mite *Neoseiulus pseudolongispinosus* (Xin, Liang and Ke) (Phytoseiidae). Biological Control. 2013;**65**(1):37-42

[121] Mullen GR, OConnor BM. Mites. In: Mullen G, Mullen GR, Durden L, editors. Medical and Veterinary Entomology. 2nd ed. Academic Press; 2009. pp. 423-482

[122] Canfield MS, Wrenn WJ. *Tyrophagus putrescentiae* mites grown in dog food cultures and the effect mould growth has on mite survival and reproduction. Veterinary Dermatology. 2010;**21**(1):58-63

[123] Jeong KY, Kim WK, Lee JS, Lee J, Lee IY, Kim KE, et al. Immunoglobulin E reactivity of recombinant allergen Tyr p 13 from *Tyrophagus putrescentiae* homologous to fatty acid binding protein. Clinical and Diagnostic Laboratory Immunology. 2005;**12**(5):581-585

[124] Malik A, Gulati R, Duhan K, Poonia A. *Tyrophagus putrescentiae* (Schrank) (Acari: Acaridae) as a pest of grains: A review. Journal of Entomology and Zoology Studies. 2018;**6**(2):2543-2550

[125] Mostafa AM, Sakr HHI, Yassin EMA, Abdel-Khalik AR. Effect of different diets on the biology of the Astigmatid mite *Tyrophagus putrescentiae*. Egyptian Journal of Agricultural Research. 2013;**91**(4):1439-1445

[126] Klimov PB, OConnor BM. *Acarus putrescentiae* Schrank, 1781 (currently *Tyrophagus putrescentiae*; Acariformes, Acaridae): Proposed conservation of usage by designation of a replacement neotype. The Bulletin of Zoological Nomenclature. 2010;**61**(1):24-27

[127] Nayak MK. Management of mould mite *Tyrophagus putrescentiae* (Schrank) (Acarina: Acaridae): A case study in stored animal feed. International Pest Control. 2006;**48**(3):128-130

[128] Gerson U, Weintraub PG. Mites for the control of pests in protected cultivation. Pest Management Science. 2007;**63**:658-676

[129] Maleknia B, Fathipour Y, Soufbaf M. Intraguild predation among three phytoseiid species, *Neoseiulus barkeri*, *Phytoseiulus persimilis* and *Amblyseius swirskii*. Systematic and Applied Acarology. 2016;**21**(4):417-426

[130] Beard JJ. A review of Australian *Neoseiulus* Hughes and *Typhlodromips* de Leon (Acari: Phytoseiidae: Amblyseiinae). Invertebrate Taxonomy. 2001;**15**:73-158

Chapter 9

Bioecology of Jasmine Mite, *Tetranychus urticae* in Different Jasmine Cultivars

Isaac Merlin Kamala

Abstract

Jasmine is a genus of shrubs and vines in the olive family (Oleaceae). *Jasminum sambac*, *Jasminum auriculatum*, *Jasminum grandiflorum* and *Jasminum nitidum* are the four cultivable species of *Jasminum*. The two-spotted mite, *Tetranychus urticae,* is a key pest of *Jasminum* sp. To compare the lifecycle of the notorious mite in all the cultivable *Jasminum* species, a detailed laboratory study was conducted at the Department of Agricultural Entomology, Tamil Nadu Agricultural University, Coimbatore, India. The observations on lifecycle parameters of two-spotted mite, *T. urticae,* revealed that the life cycle consists of egg, larva, protochrysalis, protonymph, deutochrysalis, deutonymph, teleochrysalis and adult. The lifecycle and duration of each stage are found to be the shortest in *J. nitidum* and the longest in *J. sambac*.

Keywords: *Jasminum sambac*, *Jasminum auriculatum*, *Jasminum grandiflorum*, *Jasminum nitidum*, two-spotted mite

1. Introduction

Flowers are inseparable from the social fabric of human life. Flowers, being adorable creation of god, befit all occasions. Jasmine is an important traditional flower, cultivated nearly throughout the tropical and subtropical parts of the world for its fragrant flowers [1]. *Jasminum sambac* is ravaged by several pests and the growers were forced to undertake frequent sprays of pesticides, with their excessive usage causing health hazards, outbreaks of secondary pests, environmental pollution, objectionable pesticide residues and adverse effect on non-target organisms and degradation of resources.

Jasminum is the generic name of shrubs and vines in the olive family (Oleaceae). Although more than 200 species are known, 40 species have been identified in India and 20 species are cultivated in South India [2–4], of which only 3 species are used for commercial cultivation namely *Jasminum sambac* (gundumalli/Madurai malli), *Jasminum auriculatum* (mullai) and *Jasminum grandiflorum* (jathimalli/pitchi). The angel jasmine, *Jasminum nitidum*, with sweetly fragrant, snow-white, pinwheel-shaped flowers, is recently introduced for commercial cultivation in Tamil Nadu, India.

1.1 *Jasminum sambac*

Jasminum sambac is an evergreen vine or shrub reaching up to 0.5–3 m (1.6–9.8 ft) tall. The flowers bloom all throughout the year and are produced in clusters of 3–12

IntechOpen

***J. sambac* plant** ***J. sambac* flower** **Flower bud**

Figure 1.
Jasminum sambac.

together at the ends of branches. They are strongly scented, with a white corolla 2–3 cm (0.79–1.18 in) in diameter with 5–9 lobes. The flowers open at night (usually around 6–8 in the evening) and close in the morning, a span of 12–20 hours. The sweet, heady fragrance of *Jasminum sambac* is its distinct feature. It is widely grown throughout the tropics from the Arabian peninsula to Southeast Asia and the Pacific Islands as an ornamental plant and for its strongly scented flowers [5] (**Figure 1**).

1.2 *Jasminum auriculatum*

Jasminum auriculatum is a species of jasmine, in the family Oleaceae. It is found in India, Nepal, Sri Lanka, Bhutan and the Andaman Islands. Due to essential oil contained in the flowers, it is cultivated commercially in India and Thailand. It is used for decorative purposes and festivals in India. It is a stunning, small climbing bushy plant with simple ovate dark green small leaves and powdery satin white flowers. Leaves are opposite, ashy-velvety, sometimes hairless, simple or trifoliolate. Lateral leaflets are much smaller, rarely exceeding 4 mm in diameter, the central one up to 3.5 cm long and 1.5 cm broad, ovate, shortly pointed. Nerves are few, lowest oblique. Bracts are linear, 4 mm long. Flowers are fragrant, in many-flowered cymes. Flower-stalks are up to 5 mm long. Calyx 3 mm long, pubescent, teeth minute. Flowers are white, tube 1.5 cm long, lobes elliptic, up to 8 mm long. Berry is 5 mm in diameter, globose, and black [6]) (**Figure 2**).

1.3 *Jasminum grandiflorum*

Jasminum grandiflorum, also known variously as the Spanish jasmine, Royal jasmine, and Catalan jasmine, among others, is a scrambling deciduous shrub growing to 2–4 m tall. The leaves are opposite, 5–12 cm long, pinnate with 5–11 leaflets. The flowers are produced in open cymes; the individual flowers are white

***J. auriculatum* plant** ***J. auriculatum* flower** **Flower bud**

Figure 2.
Jasminum auriculatum.

***J. grandiflorum* plant** ***J. grandiflorum* flower** **Flower bud**

Figure 3.
Jasminum grandiflorum.

having corolla with a basal tube 13–25 mm long and five lobes 13–22 mm long. The flower's fragrance is unique and sweet. It is widely cultivated as an ornamental plant in warm temperate and subtropical regions. By the method of solvent extraction, the Jasmine flowers are converted into jasmine concrete and jasmine oleoresin (sold as Jasmine Absolute). Both products have a huge demand in the fragrance industry. Methyl jasmonate isolated from the jasmine oil of *Jasminum grandiflorum* led to the discovery of the molecular structure of the jasmonate plant hormones [7] (**Figure 3**).

1.4 *Jasminum nitidum*

This twinning climber has slender stems and forms a dense habit. The dark glossy green leaves are lanced-shaped and the fragrant white star-shaped flowers appear throughout summer in clusters. The angel wing jasmine is grown for its flowers and glossy foliage. It is planted in small or large gardens for its fragrance or to grow over lattice or a pergola. It is suitable for coastal subtropical regions and establishes in 1–2 years. It is also used in containers and grown under glass in cold climates. Once established, it has a high-water requirement, and responds to occasional deep watering particularly during dry periods [8] (**Figure 4**).

The production of jasmine is affected by various factors, among which, insect pests are the most devasting factor. The major pests affecting jasmine are jasmine bud worm (*Hendecasis duplifascialis* Hampson), leaf webworm (*Nausinoe geometralis* Gurnee), gallery worm (*Elasmopalpus jasminophagus* Hampson.), leaf roller, (*Glyphodes unionalis* Hubner), and the two-spotted mite (*Tetranychus urticae* Koch.). Of these, budworm and two-spotted mite gain major economic importance, as they cause excessive damage to the buds and leaves, respectively. Of these, two-spotted mite has achieved the status of a major pest causing severe economic loss by reducing the vitality of the plants, thereby reducing the productivity of flowers. *Tetranychus urticae* (Koch, 1836) commonly known as two-spotted mite is a species

***J. nitidum* plant** ***J .nitidum* flower** **Flower bud**

Figure 4.
Jasminum nitidum.

Figure 5.
Symptoms of damage of two-spotted mite, Tetranychus urticae.

of plant-feeding mite that is generally considered as a pest. It is the most widely known member of the family Tetranychidae or spider mites. Spider mites are named so because many members of this family produce silk webbing on the host plants. In jasmine, flowering commences during March–April and comes to peak in May–July. During this period, the weather is too hot and is favourable for multiplication and so the population increases rapidly. These tiny eight-legged arthropods lay eggs on the underside of leaves. An adult female can lay more than 100 eggs in 3 weeks. Eggs hatch in 4–5 days and the entire lifecycle from egg to adult is completed in 1–3 weeks, depending on the temperature. The life-cycle of *T. urticae* consists of five different stages such as egg, larva, protonymph, deutonymph and adult. Mites are typically found on the underside of leaves, but may colonise entire plants during outbreaks. The mites suck sap from cells on the underside of plant leaves, in the early stages, and characteristic white speckles can be seen from the upper leaf surface. As mite number increases, these white speckles also increase and the leaf exhibits a bleached appearance [9].

In case of severe infestation, the whole plant becomes pale in colour, and affects production and size of the flower buds. Damage to the leaves inhibits photosynthesis, and severe infestations can result in premature leaf fall, shoot dieback, and decreased plant vigor. Although the individual lesions are very small, attack by hundreds or thousands of spider mites can cause thousands of lesions and thus can significantly reduce the photosynthetic capability of plants [10]. Such buds fetch a low market price. Silk webbing on the undersides of leaves is a characteristic sign of spider mites. Under high population densities, the mites move to the tip of the leaf or top of the plant and congregate using strands of silk to form a ball-like mass, which will be blown by winds to new leaves or plants, in a process known as "ballooning' (**Figure 5**).

As the infestation by the two-spotted mite, *T. urticae,* and the jasmine leaf webber, *Nausinoe geometralis* Guenee, coincides with the flushing stage, the silky foliage of jasmine is severely affected, and thereby, the photosynthetic efficiency of plant is affected, hence affecting flower production. The rapid developmental rate, short generation time, and high net reproductive rate of *T. urticae* allow them to achieve damaging population levels very quickly when growth conditions are suitable, resulting in an equally rapid decline of host plant quality.

Management of this pest has become a menace for the jasmine growers. The knowledge on life history of the pest as well as the life-table on different varieties is essential for developing IPM in better management of any pests. The knowledge of the sequence of developmental stages, their duration, and number of generations and method of overwintering is essential to know the 'weakest link' in the lifecycle. This would help to aim control measures effectively at the most vulnerable stage of the pest. The biology of jasmine two-spotted mite is attempted in the four cultivable jasmine species, that is, *Jasminum sambac*, *J. auriculatum*, *J. grandiflorum* and *J. nitidum*. The objective of the study is to compare the different life stages of two-spotted mite in different Jasmine species.

2. Materials and methods

The life history of red spider mite was studied under laboratory conditions at the Department of Agricultural Entomology, Tamil Nadu Agricultural University, Coimbatore, India with the prevailing weather parameters. Pure culture of red spider mite, *Tetranychus urticae*, was initiated by collecting the adults from jasmine field and was maintained in the laboratory. The mites were reared on jasmine leaves following the technique suggested by Rodriguez [11] and Gilstrap [12]. Field-collected mites from the field-infested jasmine leaves were reared continuously on the leaf discs of fresh mulberry leaves cut into squares of 8 cm^2 size placed on foam pads on plastic trays facing the basal side upward. Distilled water was used to keep the foam pad wet and to maintain the leaf discs in turgid conditions which were changed when they started turning yellow. Further, in order to prevent the migration of *T. urticae* colonies from one disc to other, the leaf squares were fenced with wet cotton threads. Cut infested jasmine leaf samples were first examined for predators (*i.e.*, predatory thrips or mites) to avoid possible contamination of culture. The leaves were then laid on top of the clean mulberry leaves (**Figure 6**). After establishment of the culture, the biology was studied.

About 20 mated females were released on a fresh leaf in the petridish and allowed overnight, in order to obtain the eggs. Next morning, the eggs were carefully lifted with the help of a moistened 00 size camel hairbrush and transferred

Figure 6.
Mass culturing of Tetranychus urticae.

to previously prepared leaf discs at the rate of one egg per leaf disc per Petri plate. Such 35 plants were maintained to study the biology. The development of various stages of the mite was observed twice a day with the help of stereoscopic binocular microscope. The observation on life history included incubation period (days), duration of larva, (days) protonymph, deutonymph, quiescent stages, preoviposition, oviposition and post oviposition periods, fecundity and longevity of adults, sex ratio and the viability of eggs. The midpoint between two observations was considered as the time of moulting whenever a change to next stage was observed. But, if moulting was just taking place at the time of observation, then it was taken as the time of moulting. When the mites reached the adult stage, the other sex of the mite that developed on the leaf disc was released on to the leaf disc to observe mating, pre-oviposition, oviposition and post-oviposition and also fecundity and longevity of the mite.

To study the biology in different *Jasminum* species, the mites were cultured in *J. sambac*, *J. auriculatum*, *J. grandiflorum* and *J. nitidum* placed inside Petri dishes, and the following parameters were recorded.

2.1 Fecundity rate

The female deutonymphs and male adults were collected from the respective cultures and released at one pair per Petri dish. They were allowed to oviposit, and observations on the number of eggs laid per day, egg period and total number of eggs laid throughout the oviposition period were recorded.

2.2 Duration of life stages

One pair of matured female and male mites was released per Petri dish. After a few hours of oviposition, the mites were removed using a camel hair brush retaining only five eggs in each petri dish. Hatching was observed, and the duration of different immature stages followed by adult longevity was recorded.

2.3 Percentage of larvae becoming adult

The female deutonymphs and adult males were collected from the respective cultures and released at three pairs per petri dish. Two days after oviposition, mites were removed leaving only the eggs. The numbers of larvae successfully reaching the adult stage were counted in each petri dish and the percentage was worked out.

3. Results and discussion

The biology of two-spotted mite, *T. urticae* on *J. sambac*, *J. auriculatum*, *J. grandiflorum* and *J. nitidum* as host plants was investigated under controlled conditions, and the results are presented in **Table 1**. The different life stages are described hereunder.

3.1 Egg

The freshly laid eggs were brown or translucent white in colour, which gradually turned to deep brown and then creamy pinkish as they approach hatching. The spherical-shaped eggs were laid singly. The incubation period ranged from 1 to 3 days. The average incubation period was 2.2, 2.0, 1.5 and 1.5 days in *J. sambac*, *J. auriculatum*, *J. grandiflorum* and *J. nitidum*, respectively (**Figures** 7 and **8**).

The newly hatched larvae were almost spherical in shape and creamy white in colour. Two bright and prominent red spots (simple eyes) were present on the dorsal sides of the propodosomal region. The larva possessed only three pairs of legs. The larval period ranged from 2 to 3 days. The average larval period was 2.5, 2.2, 1.9 and 1.7 days in *J. sambac*, *J. auriculatum*, *J. grandiflorum* and *J. nitidum*, respectively.

3.2 Protochrysalis

The dark green matured larva entered into the quiescent stage by anchoring itself to the leaf surface. The stage was dark green. The average period of

Life stages of two-spotted mite	Duration of different life stages of *T. urticae* in four *Jasminum* species Mean ± SD (in days)			
	J. sambac	*J. auriculatum*	*J. grandiflorum*	*J. nitidum*
Incubation period	2.2 ± 0.91	2.0 ± 0.81	1.5 ± 0.84	1.5 ± 0.84
Larval period	2.5 ± 0.52	2.2 ± 0.42	1.9 ± 0.31	1.7 ± 0.483
Duration of protochrysalis	2.2 ± 0.42	1.9 ± 0.32	1.6 ± 0.52	1.20 ± 0.42
Duration of protonymph	2.4 ± 0.52	1.7 ± 0.48	1.5 ± 0.53	1.1 ± 0.32
Duration of deutochrysalis	2.5 ± 0.53	2.0 ± 0.47	1.7 ± 0.48	1.4 ± 0.52
Duration of deutonymph	2.7 ± 0.48	2.3 ± 0.48	2.0 ± 0.67	1.0 ± 0.48
Duration of teleochrysalis	2.9 ± 0.32	2.2 ± 0.63	1.5 ± 0.71	1.1 ± 0.32
Adult longevity	21.7 ± 1.64	21.3 ± 1.64	21.1 ± 2.23	19.9 ± 1.37
Total developmental period (egg-adult)	24.9 ± 1.37	24.0 ± 1.83	23.9 ± 1.37	22.8 ± 1.55

Mean of three observations.

Table 1.
Life stages of two-spotted mite, Tetranychus urticae, on four Jasminum species.

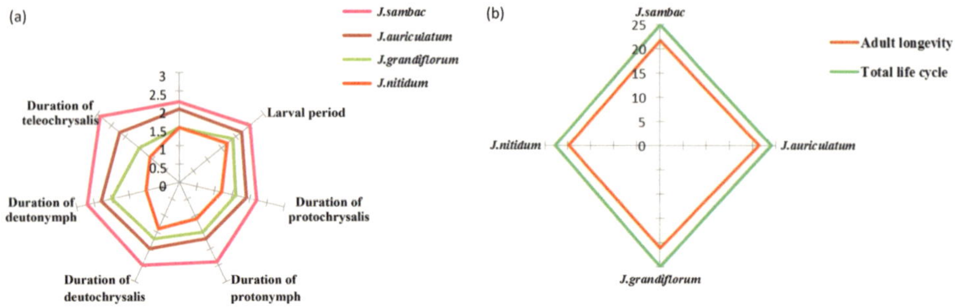

Figure 7.
Radar representation of biology of two-spotted mite, Tetranychus urticae on different Jasminum species. (a) Egg and larval period, (b) Adult longevity and total lifecycle.

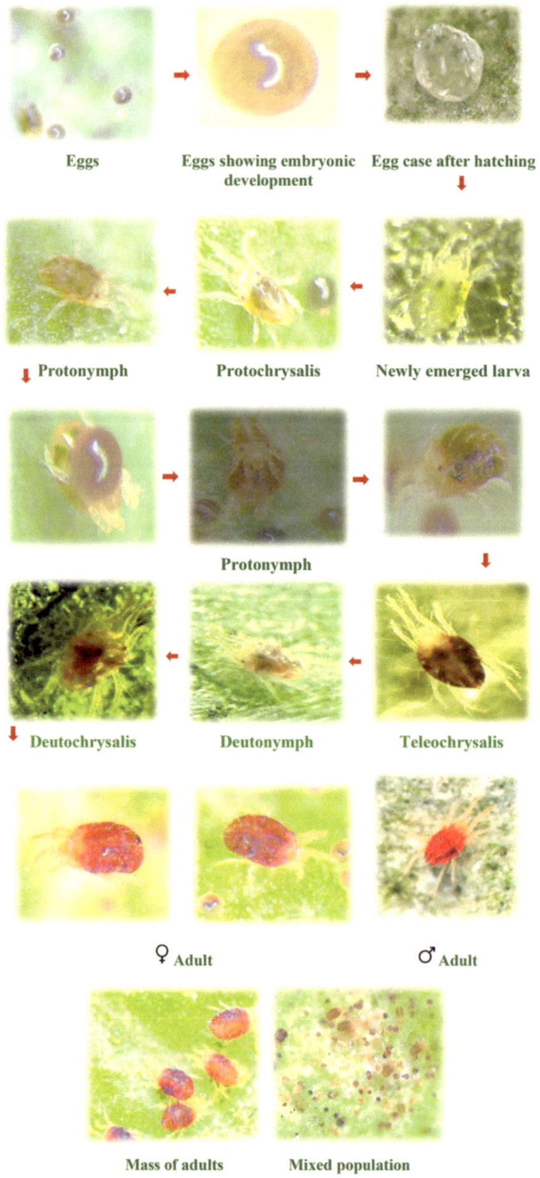

Figure 8.
Biology of two-spotted mite, Tetranychus urticae.

protochrysalis was 2.2, 1.9, 1.6 and 1.2 days in *J. sambac*, *J. auriculatum*, *J. grandiflorum* and *J. nitidum*, respectively.

3.3 Protonymph

The protochrysalis moulted into protonymph. The body was oval in shape with four pairs of legs and dark green in colour in the beginning, which later turned into amber colour. The average period of protonymph was 2.4, 1.7, 1.5 and 1.1 days in *J. sambac*, *J. auriculatum*, *J. grandiflorum* and *J. nitidum*, respectively.

3.4 Deutochrysalis

The matured protonymph entered into quiescent, a stage which is known as deutochrysalis. The body also shrinked and decreased in size and attained a dark green colour. The average period of protochrysalis was 2.5, 2.0, 1.7 and 1.4 days in *J. sambac*, *J. auriculatum*, *J. grandiflorum* and *J. nitidum*, respectively.

3.5 Deutonymph

The deutonymph emerged from deutochrysalis. The body was red coloured, larger and broader than protonymph. The average period of deutonymph was 2.7, 2.3, 2.0 and 1.0 days in *J. sambac*, *J. auriculatum*, *J. grandiflorum* and *J. nitidum*.

3.6 Teleochrysalis

The deutonymph at its maturity enters into a quiescent stage known as teleochrysalis. In this stage, the body shrinks and decreases in size. The colour of this stage is light red to creamy. The average larval period of protochrysalis was 2.9, 2.2, 1.5 and 1.1 days in *J. sambac*, *J. auriculatum*, *J. grandiflorum* and *J. nitidum*.

3.7 Adult

The body of the adult male was narrow with a distinct abdomen, greenish in colour which later turned pinkish. The first pair of legs was longer than the rest of the pairs. Males were smaller than females and lived for 11–12 days. The newly emerged females looked dull red which later turned to deep brick red. The simple eyes were seen as two red spots on the sides of the dorsal propodosomal region. The adult female survived for 19–24 days. The pre-oviposition period varied from 2 to 3 days with an average of 2.7 days. The oviposition period lasted for 11–14 days with an average of 12.0 days. Each female laid about 123–160 eggs with an average of 146.6 eggs. The mean number of eggs laid was 10–12 eggs per day. The maximum number of eggs laid by a female was 11–12 eggs per day. The average adult longevity is 21.7, 21.3, 21.1 and 19.9 in *J. sambac*, *J. auriculatum*, *J. grandiflorum* and *J. nitidum*, respectively.

3.8 Total developmental period

The total developmental period (egg to adult) ranged from 24.9, 24.0, 23.9 and 22.8 days in *J. sambac*, *J. auriculatum*, *J. grandiflorum* and *J. nitidum*, respectively (**Figure 6**).

The biology of the two-spotted mite, comprising egg, larva, protochrysalis, protonymph, deutochrysalis, deutonymph, teleochrysalis and adult stages was completed in 24.9 days. The results are in agreement with Rajakumar et al. [13], who reported that the total developmental period of mites in jasmine was 22.80 days for

♂ and 31.08 days for ♀. The outcome of the study was also in conformity with Vinoth Kumar et al. [14] who also suggested that the developmental period ranges from 26 to 27 days in brinjal and Premalatha [15] who stated that the total developmental period of two-spotted mites lasted from 24.71 to 25.71 days. But the egg-to-adult developmental period was lesser *viz*., 24.0, 23.9 and 22.8 days in other *Jasminum* sp., *J. auriculatum*, *J. grandiflorum* and *J. nitidum*, respectively, which proves that mites did not prefer them as *J. sambac*. The biochemical contents of the plant like high phenols, low sugar, protein or other biophysical factors like leaf surface wax could be the probable reason for the less preference which resulted in shorter life cycle. Different leaf characters like leaf area, leaf hair density, length of leaf and leaf thickness have a significant impact on the searching capability of mites [16]. Saber and Momen [17] reported that leaf toughness and thickness are very important factors, which influence the reproduction and development of phytoseiid mite population.

4. Conclusion

The observations on life cycle parameters of two-spotted mite, *T. urticae,* revealed that the life cycle consists of egg, larva, protochrysalis, protonymph, deutochrysalis, deutonymph, teleochrysalis and adult. The lifecycle and duration of each stage are found to be the shortest in *J. nitidum* and the longest in *J. sambac*.

The use of moderately resistant varieties as a part of the IPM strategy can enhance the biological and chemical tools of insect pest management. One of the techniques to decrease the pest damage is use of the cultivars or species which show higher resistance to insect pests such as *J. nitidum* and *J. grandiflorum* in areas where the incidence of jasmine mites is high. *J. grandiflorum* is a commercially cultivated species of jasmine, well known for its mesmerizing fragrance as well as usage for concrete recovery. But, *J. nitidum* is a new *Jasminum* species with star-shaped fragrant flowers newly introduced for commercial flower cultivation. Hence, utilizing this least preferred species in breeding programs of jasmine will yield better varieties with greater degree of resistance to jasmine pests.

Author details

Isaac Merlin Kamala
Tamil Nadu Agricultural University, Coimbatore, India

*Address all correspondence to: merlinento@gmail.com

References

[1] Mabberley DJ. Mabberley's Plant-Book: A Portable Dictionary of Plants, their Classification and Uses. 3rd ed. New York: Cambridge University Press; 2008. pp. 35-36

[2] Bhattacharjee SK. Native jasmine of India. Indian Perfumes. 1980;**24**(3):126-133

[3] Lakshmi J, Ganga M. Floral biology studies in certain lesser known species of jasmine (*Jasminum* spp.). International Journal of Current Microbiology and Applied Sciences. 2017;**6**(8):2811-2815

[4] A Jain, R Sharma, A Kumar, S Sharma. Jasminum species: An overview. International Journal of Institutional Pharmacy and Life Sciences. 2011;(1):251-256. https://www.researchgate.net/publication/292575441_Jasminum_species_an_overview.

[5] Kumari G, Wong KH, Serra A, Shin J, Yoon HS, Sze SK, et al. Molecular diversity and function of jasmintides from *Jasminum sambac*. BMC Plant Biology. 2018;**18**:144

[6] Manimaran P, Ganga M, Kannan M. Manipulation of petal senescence in *Jasminum nitidum* flowers with packaging and pre-treatment during storage: Role of phenolics. Journal of Pharmacognosy and Phytochemistry. 2018;7(2):3478-3480

[7] Sandeep S, Paarakh P. *Jasminum grandiflorum* Linn (Chameli): Ethnobotany, phytochemistry and pharmacology—A review. Pharmacology. Pharmacologyonline 2 Newsletter. 2009;(2):586-595. https://www.researchgate.net/publication/215517134_Jasminum_grandiflorum_Linn_Chameli_Ethnobotany_Phytochemistry_and_Pharmacology_-_A_review [Accessed: 02 April 2019]

[8] Bahuguna Y, Singh RM, Juyal V, Gupt V. Antilithiatic effect of flowers of *Jasminum auriculatum* Vahl. International Journal of Green Pharmacy. 2009;(45):654-672. DOI: 10.4103/0973-8258.54910

[9] Sadeghi E, Vafaei R, Madani H. The influence of *Tetranychus urticae* Koch (Acari: Tetranychidae). Life table and reproductive parameters by applying Si on bean at library condition. Advances in Entomology. 2016;**4**:260-267

[10] Zhang ZQ. Mites of Greenhouses: Identification, Biology and Control. Cambridge, UK: CABI Publishing; 2003. p. 244

[11] Rodriquez J. Detached leaf culture in mite nutrition studies. Journal of Economic Entomology. 1953;**46**:713

[12] Gilstrap FE. Table top production of *Tetranychus urticae* mites and their phytoseiid natural enemies. Journal of Kangas Entomological Society. 1977;**50**:229-233

[13] Rajakumar E, Hugar PS, Kattimani KN. Biology of red spider mite, *Tetranychus urticae* Koch (Acari: Tetranychidae) on jasmine. Karnataka. The Journal of Agricultural Science. 2005;**18**(1):147-149

[14] Vinoth Kumar S, Chinniah C, Muthiah C, Sadasakthi A. Field evaluation of certain newer acaricide/insecticide molecules for their bioefficacy against *Tetranychus urticae* Koch on brinjal. Karnataka. The Journal of Agricultural Science. 2009;**22**(3):705-706

[15] Premalatha K. Studies on seasonal incidence and integrated management of mite complex on tomato. *Lycopersicon esculentum*. Mill. M. Sc (Ag.,) thesis. Coimbatore: Tamil Nadu Agricultural University; 2015

[16] Cedola CV, Sanchez NE, Liljesthron GG. Effect of tomato leaf hairiness on functional and numerical response of *Neoseiulus californicus* (Phytoseiidae: Acarina). Experimental & Applied Acarology. 2001;**25**(10-11):819-831

[17] Saber SA, Momen FM. Influence of plant leaf texture on the development, reproduction and life table parameters of the predacious mite *Cydnoseius zaheri* (Phytoseiidae: Acarina). Acta Phytopathologica et Entomologica Hungarica. 2005;**40**(1-2):177-184